John Croumbie Brown

Forestry in Norway

With notices of the physical geography of the country

John Croumbie Brown

Forestry in Norway
With notices of the physical geography of the country

ISBN/EAN: 9783743337404

Manufactured in Europe, USA, Canada, Australia, Japa

Cover: Foto ©berggeist007 / pixelio.de

Manufactured and distributed by brebook publishing software (www.brebook.com)

John Croumbie Brown

Forestry in Norway

FORESTRY IN NORWAY:

WITH

NOTICES OF THE PHYSICAL GEOGRAPHY OF THE COUNTRY.

COMPILED BY

JOHN CROUMBIE BROWN, LL.D.,

Formerly Lecturer on Botany in University and King's College, Aberdeen; subsequently Colonial Botanist at Cape of Good Hope, and Professor of Botany in the South African College, Capetown; Fellow of the Linnean Society; Fellow of the Royal Geographical Society; and Honorary Vice-President of the African Institute of Paris.

EDINBURGH:
OLIVER AND BOYD, TWEEDDALE COURT.
LONDON: SIMPKIN, MARSHALL, & CO.,
AND WILLIAM RIDER & SON.
MONTREAL: DAWSON BROTHERS.

1884.

ADVERTISEMENT.

───o───

IN the Spring of 1877, while measures were being taken for the formation of an Arboretum in Edinburgh, I issued a Pamphlet entitled *The Schools of Forestry in Europe: a Plea for the Creation of a School of Forestry in Connection with the Arboretum at Edinburgh.* After it was made known that arrangements were being carried out for the formation of an International Exhibition of forest products, and other objects of interest connected with forestry, in Edinburgh, with a view to promoting the movement for the establishment of a National School of Forestry in Scotland, and with a view of furthering and stimulating a greater improvement in the scientific management of woods in Scotland, and the sister countries, which has manifested itself during recent years, the Council of the East Lothian Naturalists' Club resolved on having a course of lectures or popular readings on some subject connected with forestry, which might enable them and others better to profit by visits to the projected Exhibition, and which should be open to the public at a moderate charge. The conducting of these was devolved upon me, who happened to be Vice-President of the Club,

ADVERTISEMENT.

The following treatise was compiled from information then in my possession, or within my reach; and it constituted the basis of the lectures. I had a mixed audience of ladies and gentlemen, tradesmen and young men and maidens still pursuing their studies. The technical information was laid upon the table for consultation by members of the Club or others at their convenience; the other portions only were read. It is published as a sequel which may be found useful as an introduction to another volume entitled *Introduction to the Study of Modern Forest Economy*, which was lately published in accordance with a resolution formally adopted by promoters of the Exhibition to give their best efforts and endeavours to render the Exhibition a success.

In Norwegian names I have followed, I consider, the principles of English orthography, but I have not altered the orthography adopted in passages by others which I have so largely quoted.

JOHN C. BROWN.

HADDINGTON, *1st July, 1884*.

CONTENTS.

———o———

CHAPTER I.—*General Features of the Country,* . 1-8
Description of Northern Scandinavia, by M. du Chaillu (p. 1); Christiania and Torristal Rivers (p. 3); Skager Rack and Christiania Fiord (p. 6).

CHAPTER II.—*Forest Scenes,* 9-22
The Glommen and Floating Timber (p. 9); Description of Forests by M. Guillemard (p. 11); Descriptions by Derwent Conway (p. 15); Thomas Forester (p. 15); and Edward Price (p. 22).

CHAPTER III.—*Mountain Plateaux and Mountain Ravines,* 23-36
Christiania to Vest Fiord (p. 23); The Marie Stege (p. 25); The Rinkan Foss (p. 29).

CHAPTER IV.—*Geographical Distribution of Trees in Norway,* 37-49
Reports on the Subject (p. 37); Details by Dr Broch (p. 39); by M. Marny (p. 48).

CHAPTER V.—*Conditions on which depend the Distribution of Vegetation,* 50-57
Views of Forbes (p. 50); Schow (p. 51); Illustration by Schleiden (p. 54).

CHAPTER VI.—*Conditions Affecting the Geographical Distribution of Plants and Trees in Norway,* . 58-62

CONTENTS.

CHAPTER VII.—*Temperature*, ... 63-72
Illustrations by Schleiden (p. 64); Meteorological Observations in Norway (p. 67); Isothermal Lines (p. 69).

CHAPTER VIII.—*Rainfall and Moisture*, . . 73-81
Rain (p. 73); Moisture (p. 77); Foggy Days (p. 80).

CHAPTER IX.—*Rivers*, ... 82-88
Natural History of Rivers (p. 82); Notices of Rivers by Dr Broch (85).

CHAPTER X.—*Lakes*, ... 89-96
The Miosen Lake (89); Notice of George P. Bidder (90); Depths of Lakes (94).

CHAPTER XI.—*Winds*, ... 97-102
Thunder Storms (p. 98); Storms of Wind (99).

CHAPTER XII.—*Geological Formations*, . 103-113
Limitation of Habitat of Plants by Composition of the Soil (p. 104); Geological Formation in Norway (p. 108).

CHAPTER XIII.—*Mountains and Fjelds*, . 114-134
Affect of Altitude on different Species of Plants (p. 114); Barometric Pressure of Atmosphere in Norway (p. 121); Altitude of Land (p. 123-132); Detailed Information in regard to the Mountain Systems of Norway (p. 124); Characters of Fjelds or High-lying Plateaux (p. 129).

CHAPTER XIV.— *Temperature and Altitudes of Snow-Fields and Glaciers*, . . . 135-137
Characteristics (p. 134), and Altitudes of these (p. 136).

CHAPTER XV.—*Mechanical Action of Glaciers*, 138-163
Description of Sogne Fiord (p. 139); View from the Sogndal (p. 143); Waterfalls and High land behind (p. 144); Remarkable Depths in the Fiord (p. 145); Indications of Glacial Action (p. 146); *Striae* and Moraines (p. 157).

CONTENTS.

PAGE

CHAPTER XVI.—*Appearances Presented by Glaciers and Snow-Fields,* 164-175

The Justedal Glacier (p. 164); Motion of a Glacier (p. 166); The Ringedal Waterfall (p. 168); Description of Veblungsnaesset (p. 171); The Stiggevand (p. 174).

CHAPTER XVII.—*Saeter Life,* . . . 176-189

Description of the Nisservand (p. 176); Account of Saeters or Mountain Cheese Farms (p. 178); Saeter in the Valdal (p. 179); Roeldal Saeter (p. 182).

CHAPTER XVIII.—*Valleys,* . . . 190-194

Valleys in the Guldbranbsdal (p. 190).

CHAPTER XIX.—*Forest Exploitation, and the Transport of Timber, and Export Timber Trade,* 195-203

Exploitation Determined to some Extent by the Land Tenure (p. 195); Glissoirs or Artificial Shoots (p. 196); Floatage (p. 198); Export Timber Trade (p. 200).

CHAPTER XX.—*Shipbuilding and Shipping,* . 204-211

Type of Boat Maintained (p. 204); Ancient Boats (p. 205); Modern Shipping (p. 207).

CHAPTER XXI.—*Forest Devastation,* . . 212-219

Illustration Supplied by State of Forests in Romsdal (p. 212), and in Prefectures of Lister and Mandal (p. 218).

CHAPTER XXII.—*Remedial Measures,* . . 220-227

Sylviculture at Agricultural School at Aas (p. 227); Staff of Officials in the Forest Service of the Government (p. 227).

AUTHORITIES CITED.

AARS, p. 38; ANDREWS, p. 41; BALFOUR, pp. 69, 114, 118; BARTH, p. 38; BROCH, pp. 37, 39, 40, 43, 58, 69, 73, 76, 83, 96, 98, 108, 121, 161, 190, 198, 207; VAN BUCH, p. 137; *Building News*, p. 49; AN OLD BUSHMAN, p. 134; *Budget Report*, p. 225; CAMPBELL, p. 131; DU CHAILLU, pp. 1, 23, 139, 164, 179; ESMARK, p. 154; *Ecclesiastes*, p. 170; FORBES, pp. 83, 124, 134, 155, 161; FORESTER, pp. 15, 198; *Forest Lands of Finland*, p. 148; *Forests and Moisture*, pp. 74, 151; FOUNGER, p. 219; *Frost and Fire*, pp. 5, 154, 177, 198; GEIKIE, pp. 149-150; GUILLEMARD, p. 11; *Hamburg Correspondent*, p. 205; HEER, p. 62; HERSHEL, p. 182; HULLET, p. 39; *Hydrology of South Africa*, p. 75; LOUDON, p. 39; MADDEN and STRACHY, p. 117; MARNEY, p. 48; MEYER, p. 120;

FORESTRY OF NORWAY.

CHAPTER I.

GENERAL FEATURES OF THE COUNTRY.

M. DU CHALLU, in his volume entitled *The Land of the Midnight Sun*, writes:—'There is a beautiful country far away towards the icy north. It is a glorious land; with snowy, bold, and magnificent mountains; deep, narrow, and well-wooded valleys; bleak plateaux and slopes; wild ravines; clear and picturesque lakes; immense forests of birch, pine, and fir trees, the solitude of which seems to soothe the restless spirit of man; large and superb glaciers, unrivalled elsewhere in Europe for size; arms of the sea, called fjords, of extreme beauty, reaching far inland in the midst of grand scenery; numberless rivulets, whose crystal waters vary in shade and colour as the rays of the sun strike upon them on their journey towards the ocean, tumbling in countless cascades and rapids, filling the air with the music of their fall; rivers and streams which, in their hurried course from the heights above to the chasm below, plunge in grand waterfalls, so beautiful, white, and chaste, that the beholder never tires of looking at them; they appear like an enchanting vision before him, in the reality of which he can hardly believe. Contrasted with these are immense areas of desolate and barren land and rocks, often covered with boulders which in many places are piled here and there in thick masses, and swamps and moorlands, all so dreary, that they impress the stranger with a feeling of loneliness from which he tries in vain to escape. There are also many exquisite sylvan landscapes, so quiet, so picturesque, by the sea and lakes, by the hills and the mountain sides, by the rivers and in the glades, that one delights to linger among them. Large and small tracts of cultivated land or fruitful glens, and valleys, bounded by woods or rocks, with farm-houses and cottages, around which fair-haired children play, present a striking picture of contentment. Such are the characteristic features of Scandinavia, surrounded almost everywhere by a wild and austere coast. Nature in Norway is far bolder and more majestic than in Sweden; but certain parts of the coast along the Baltic present charming views of rural landscape.

'From the last days of May to the end of July, in the northern part of this land, the sun shines day and night upon its mountains, fjords, rivers, forests, valleys, towns, villages, hamlets, fields, and farms; and thus Sweden and Norway may be called "The Land of the Midnight Sun." During this period of continuous daylight the stars are never seen, the moon appears pale, and sheds no light upon the earth. Summer is short, giving just time enough for the wild-flowers to grow, to bloom, and to fade away, and barely time for the husbandman to collect his harvest, which, however, is sometimes nipped by a summer frost. A few weeks after the midnight sun has passed, the hours of sunshine shorten rapidly, and by the middle of August the air becomes chilly and the nights colder, although during the day the sun is warm. Then the grass turns yellow, the leaves change their colour, and wither and fall; the swallows, and other migrating birds, fly towards the south; twilight comes once more; the stars, one by one, make their appearance, shining brightly in the pale blue sky; the moon shows itself again as the Queen of Night, and lights and cheers the long and dark days of the Scandinavian winter. The time comes at last when the sun disappears entirely from sight; the heavens appear in a blaze of light and glory, and the stars and the moon pale before the aurora borealis.'

GENERAL FEATURES OF THE COUNTRY.

It is now some years since I visited Norway and Sweden; but I still keep up communication with friends interested in forestry in both countries. I was at that time going to St. Petersburg to preach in the British and American chapel, while the minister sought a little relaxation at home. I commenced my ministry in that city. I had frequently acted in like manner as the *locum tenens* of others, my successors, there. On most occasions of my doing so I have made the journey by different routes; and on several of these I have given attention to the forestry of the countries through which I passed. By a desire to see something of the forests and forestry of Norway and Sweden, my route on this occasion was determined.

I sailed from Leith for Copenhagen. We had a rough passage across the German Ocean. We reached Christiansand at six o'clock in the morning; and I determined to disembark. A steamer for Christiania had sailed shortly before our arrival, but there was another in port which was to sail the following morning, and thus I had the day at my disposal. I walked first to the cemetery, and after breakfast, returning thither, I rambled about on the Lovers' Walk—a walk adjacent, extending over miles of well-kept footpaths, winding about interminably upon rocks and rising ground to the left of the land as you approach by sea; and I found some of the views of the fiord, and of adjacent fiords, dotted with islands, exceedingly beautiful. After dinner I took a sail by a small steamer up the Torristal river, returning by seven P.M. This little voyage, a similar one up the Topptal river, and one which would have taken me out in the fiords, were all recommended to me. My choice was made capriciously. I believe, from what I was told, that I would have found either of the others equally pleasant, but different. I found my morning ramble and my afternoon trip together a good preparation for visiting more extensively the beautiful and romantic scenery of Norway.

On my return voyage on the Torristal river I met with

a timber merchant, who had that day completed the purchase of a ship-load of wood which was still growing in the forest; and who courteously and frankly informed me in regard to the transaction. The farmer, a small landed proprietor, had engaged, on terms agreed to, to fell, prune, and deliver in the river the quantity of wood required, in logs of a specified average length and girth. All the *débris* remained his property, and might be cut up for firewood and used or sold, or otherwise disposed of as he might please. And the logs would be, when delivered and accepted, floated down to Christiansand, and shipped by the purchaser.

I have used the expression farmer or small landed proprietor, for such is the character of most of the inhabitants of this district. The banks of the Topptal river are dotted with houses, and with mansions, which speak of greater wealth or more extensive possessions; and many of the farmers are dependent on the cutting and sale of wood to enable them to pay their rent. The logs I found to be of but limited dimensions, and I asked the timber merchant if the exploitation adopted did not tend to devastate the woods. 'Look,' said he in reply, 'there are young trees growing in every corner down to the water's edge on both sides of the river, throughout the whole course of our voyage. And so is it for miles inland. As fast as we fell others grow.' I called his attention to the small size of the logs, and told him of what was being done in Sweden and elsewhere for the conservation and improvement of forests; but he only laughed, as if that were a thing altogether unnecessary here, and one which it would be ridiculous to propose, and he called my attention to the floats of timber we were constantly passing.

At one mill at Christiansand it is said 70,000 trees are thus floated down and sawn up every year, and there are several other saw-mills in the town. At Vigelund, about ten miles above the town on the Torristal river, a little way above the rapids of the river to which the steamer goes, there is another saw-mill. At the fall on which this mill stands may be seen what I shall hereafter have occasion to refer to as characteristic of the water transport of timber trees in Norway. The author of a volume entitled *Frost and Fire*, describing the passage of this fall by trees, says :—' At every moment some new arrival comes sailing down the rapids, pitches over the fall, and dives into a foaming ground pool, where hundreds of other logs are revolving and whirling about each other in creamy froth. The new comer first takes a header, and dives into some unknown depth, but presently he shoots up in the midst of the pool, rolls over and over, and shakes himself till he finds his level, and then he joins the dance. There is first a slow sober glissade eastward across the stream to a rock which bears the mark of many a hard blow. There is a shuffle, a concussion, and a retreat, followed by a pirouette sunwise, and a sidelong sweep northwards up stream towards the fall. Then comes a vehement whirling over and over, or if a tree gets his head under the fall, there is a somersault, like a performance in the Halling dance. That is followed by a rush sideways and westward, when there is a long fit of setting to partners under the lee of a big rock; then comes a simultaneous rush southwards, towards the rapid which leads to the sea, and some logs escape and depart, but the rest appear to be seized with some freak, and away they all slide eastwards again across the stream to have another bout with the old battered pudding-stone rock below the saw-mill; and so for hours and days logs whirl one way, in this case against the sun, below the fall, and they dash against the rounded walls of the pool. Such is the effect of these concussions that above the fall it has been found necessary to protect the rock against floating bodies so as to preserve the way of the stream. It threatened to alter its course and leave the mill dry, for the rock was wearing rapidly. Lower down, nearer the sea, is a long, flat marsh, between high, rounded cliffs; and there these mountaineers, floating on to be sawn up, form themselves into a solemn funeral procession, which extends for miles; and it may be noticed that the course of this stream of floats is always longer than the course of the river's bed; for the water is slowly swinging from side to side as it flows, and the floats show the course of the stream and its whirling eddies.'

The Halling dance referred to is a Norwegian dance, reminding one at once of the sailors' hornpipe and the Highland fling, but still more vigorous and exhausting to the dancer, being diversified with somersaults such as are here alluded to.

On my return to Christiansand in the evening I at once took possession of my berth on board the steamer, which, like others in which I have sailed on these northern seas, was more luxurious in its arrangements than are those in many of the sea-going steamers on the British coast.

We sailed at three o'clock in the morning, and from that time, till near nine o'clock in the evening, with the exception of about an hour and a half in the middle of the day when we left the coast, we were the whole way sailing onwards among islands, rounding them and passing them as if on a pleasure trip—there being always rocks to seaward to break the roll of the waves, and secure for us placid waters,—and looking in on every village on the coast, while breakfast, dinner, and supper were served with all the northern whets and appetisers of which many a one has heard.

Steamers from Britain to Christiania generally avoid the coast, so that this sight is lost until they enter the Christiania Fiord. Such as is the fiord, such was the whole course, with the trifling exception of the hour and a half spoken of. I was reminded of a voyage through the Thousand Isles of Lake Ontario in America; but the scene was different. Here the islands are rocks, not rocks rough and rugged—but rocks of granite planed down and smoothed by glacial action, more like clean and white and sparkling banks of mud than are rocks on a sea-girt shore. It required no effort and but little fancy to picture them as an ocean bed rising above the sea, when, according to the Hebrew cosmogony, God said, "Let the waters of the earth be gathered together, and let the dry land appear." And again, "Let the earth bring forth grass, the herb yielding seed, and the fruit tree yielding fruit after his kind: and it was so."

There, were the bare, rounded granite rocks, without a blade of vegetation; there, were others with only a lichen, or a moss, or a grassy, or a flowery green spot. The former was on a dry rock, the latter on any crack or hollowed basin; and there, where there was a wider rent or a cup-like basin containing a handful of earth, a sapling tree; and there, an island not much larger, covered with trees to the water's edge, and there, larger islands, or the main land, with high rising hills clothed with wood and forests beyond. Again and again I felt that day as if I were alone with God, or rather with His works, as His work is described by Wisdom in the Book of Proverbs: 'When there were no depths, I was brought forth; when there were no fountains abounding with water. Before the mountains were settled, before the hills was I brought forth: while as yet He had not made the earth, nor the fields, nor the highest part of the dust of the world. When He prepared the heavens: when He set a compass upon the face of the depth: when He established the clouds above: when He strengthened the fountains of the deep: when He gave to the sea His decree, that the waters would not pass His commandment: when He appointed the foundations of the earth.'

The effect was heightened by the general absence of animal life, excepting at the towns and villages. Once or twice a cow was seen, once or twice a bird on the wing, and once a realisation of Kingsley's picture of the sea-gull on the All-alone stone far out at sea! This was as we left the islands shortly after noon. There were four gulls struggling to maintain their footing on a little projecting rock far out at sea, washed over by a wave produced by our passing vessel. And here and there a solitary house, neat painted and clean, might be seen, or an island where there was none but the highway of the sea, as if man were only beginning to appear upon the earth.

In travelling thus far one meets chiefly with a stalwart race of yeomen, presenting very much the same general appearance as do Americans in rural districts in the United States, or as do substantial Dutch boors in the inland districts of the colony of the Cape of Good Hope. But in Christiania there is a museum of Scandinavian curiosities, amongst which are life-size figures of Norwegian peasants in picturesque national costumes, which I had previously seen do good service at one of the International Exhibitions, either that at Paris in 1867, or that at Vienna in 1873—and which have been secured for permanent exhibition here. I may mention in passing that I was struck with the resemblance of many of the Norwegians of all classes, both men and women, to personal friends of my own in Scotland. I have named a dozen, and might have named scores of friends whose figure, gait, and countenance I found completely reproduced; and I have never found this in other lands.

CHAPTER II.

FOREST SCENES.

FROM Christiania I took a trip towards Drammen, and saw what Norway is under cultivation; and a journey towards Aarnaes and Charlottenburg gave me, in the first part of that journey, an opportunity of seeing under cloud and rain, Norway in a similiar condition to that of our moorland districts in Britain. As if the former trip had shown what the earth was when the earth was young—inhabited by man and cultivated, but young—this seemed to show what appearance the earth puts on in old age; and in journeying from Aarnes towards Charlottenburgh I found yet another aspect presented.

There are extensive districts in the vicinity of Glasgow, of Newcastle, and of Durham, where it appears to be coal, coal, coal, and iron and coal, but chiefly coal, which constitute there the one article of transport and products of the locality.

In America, again, in travelling through the so-called oil district lying between Pittsburg and the eastern shores of Lake Ontario, it is oil, oil, oil,—oil everywhere,—what seem interminable trains of waggons, but oil cisterns all of them, and pipes like large water pipes or drain pipes—all conveying oil.

Here it is wood, wood, wood, or perhaps I should say timber, timber and wood, everywhere. Wood by the roadside, trucks laden with wood, wood piled at the stations and on the fields, and last of all a river covered with wood and floating timber.

This is the Glommen, here a broad river, and apparently

deep; near the railway bridge by which it is crossed, the logs have been collected into floating islands of wood, begirt and confined by a chain, of which the links are logs, logs with a hole bored at either end, and tied one to another by withes. As we proceed we see the river bearing hundreds and thousands of logs onward to this gathering-place. The size of the river, compared with the size of these, suggests the idea of some boys having emptied into a brook a hundred, or a thousand, or a hundred thousand boxes of matches, and we looking on seeing them floating away. Again and again we came upon a little fall, one of three or four feet, and there the logs came tumbling down sometimes sideways, sometimes slanting, sometimes head foremost, kicking up their heels in the air.

The river is broad, it comes curving along through woodlands, these partly concealed; and I felt as if I could realise the graphic picture given by Hugh Miller of a river in pre-Adamic times bringing down the forestal products which afterwards were converted into fields of coal.

The Glommen is the principal river in Norway. It orginates in the lake Oresund, under the 62° of north latitude, and runs southward about 90 miles through a rugged channel full of cataracts and shoals. One of its confluents is the Worm, which flows through Lake Myosen. Before their confluence it is as large as the Thames at Putney, and about 20 miles below this it flows into the sea at Frederickstadt. Its highest cataract is that of Sarpen, which is 60 feet perpendicularly, and is not far from its influx into the sea.

In regard to Norwegian forests, I have heard a tourist in Norway complain that he had seen none. He had seen what I had seen of Bohus Bay and Christiania Fiord. He had visited Myosen Lake, and, if I mistake not, gone as far as Lillehammer, but he had only seen such like young woods as I have described as seen on the Torrisdalelv. I should scarcely have used the term *complain*—it was rather disappointment than complaint which was expressed, for he had perceived the fact that all forests of easy access had been subjected to exploitation, repeated again and again, so soon as the trees of the renewed forest had attained a size which gave them a marketable value, which was long before they could attain to the growth of what are known as forest trees.

Mr A. G. Guillemard, with whose graphic accounts of his Forest Rambles in many lands are enriched the pages of *Forestry*, thus tells of the forests of Norway :—
'If it were not for the broken character of the country, which is everywhere diversified by mountain ranges, undulating plains of small area, on which the peasants grow fair crops of barley, oats, and potatoes, and deep ravines, down which broad blue torrents, beloved of trout and salmon fishers, plunge madly in their seaward course, the forest scenery of Southern Norway might be deemed somewhat monotonous. There is but little variety of foliage, for Norwegian timber trees are few in number of species. Those most common are the spruce and Scotch fir, both of which attain a great size in favoured localities, both in point of altitude and girth. In many places one may observe magnificient specimens of spruce firs of from a hundred to as much as a hundred and fifty feet, in height, and of exquisite symmetry and grace. When of this great height, however, they lose some portion of their beauty of form by reason of the lower branches having decayed, and either having fallen off from the parent stem or become denuded of foliage. The Scotch firs suffer still more in this respect, and all that attain any great height are bare of branches for the lower sixty or eighty feet, rearing straight ruddy trunks crowned with wide-spreading boughs, thickly clad with dark-green needles. In the magnificent forests of pine and fir which clothe the Sierra Nevada range in California—to my mind the most beautiful forests in the world, those of tropic countries not

excepted—the trees wear a different aspect. The mountain slopes are, perhaps, rather less thickly timbered, and the trees consequently get more light, air, and breathing space; but, however this may be, the individual trees show a healthier growth and far more abundant leafage. Again and again, when riding along a dim trail in Californian forest country, I have reined in my horse to gaze in silent admiration at matchless specimens of spruce and silver fir, far exceeding a hundred feet in altitude, and yet preserving a profuse leafage from butt to summit, the lower branches, decorously sweeping the ground, being as green and luxuriant as the tiny twigs far aloft in the sunlight, and each tree being absolutely perfect in symmetry, and one of the most beautiful of Nature's creations in point of elegance of outline and tracery of foliage. Very probably the shallowness of the soil all over Norway, except in the alluvial valleys, will account for this comparative sparseness of foliage; for it is far more noticeable on the higher ground, many of the trees in the ravines and valleys, when the soil is deeper and the situation less exposed, being of very luxuriant growth. Other species of pine and fir are rarely met with, though I found the silver fir growing well in numerous clumps between Eide and Vossevangen to the north of Hardanger fiord. Beech, oak, and elm, grow fairly in places, but cannot be classed among Norweigan forest trees: they are mainly to be found near the towns, and especially in the environs of Christiania and Bergen. In Northern Norway these species are unknown. The birch is found in almost every district, as also the mountain ash, and the alder and aspen grow freely in the valleys.

'The sylvan scenery of the lightly-timbered lowland valleys can hardly be surpassed in point of variety of detail, brilliancy of colouring, and tranquil loveliness. The valley which trends north-east from Vossevangen to Stalheim, at the head of the world-famous Nærodal, affords on a fine day in autumn an exquisite series of views of this character. The road, terraced along the lower slopes of the foot-hills on the northern side, is never more than a quarter of a mile distant from the broad blue river, which for mile after mile flows through a string of picturesque lakelets whose gleaming waters reflect in vivid outline the groups of stately spruces on the tiny islands and jutting promontories. In the pure mountain air and beneath a glowing sky, whose depth of blue would almost bear comparison with that of California, the brilliant colouring of the northern landscape in early autumn, when the first night-frosts are beginning to paint the woods, must be seen to be realised. England's 'autumn gold' is charming indeed beneath an October sun, but its tints are, comparatively speaking, sober and sad. Norway's gold is bright and red and glowing, and cheers as well as charms. The sunlight streaming down through the interlacing boughs of the firs shines on the glossy dark leaves and scarlet clusters of the *moltebœre* spangling the surface of the ground in bright parterres, and lights up joyously the exquisite shades of green, ruddy-brown, and deep crimson of the mosses and the varied purple and silver grey of the lichens which clothe the masses of granite lying here and there amidst the undergrowth. Where the woodland opens out down by the river bank, the bracken is already yellowing and the purple of the heather gives signs that its beauty is on the wane. The lower leaves of the aspen, tremulous in the light air which is always breathing above the river, are turning red-brown under the influence of the chill nights; the cotton plants in the tiny swamps are waving their white plumes in token of surrender; even the hardy alder has a touch of bronze on its foliage, and the mountain ash is everywhere gorgeous with its pendulous clusters of crimson berries. Up the slopes of the hills, amongst the dark green of the spruces, and the ruddy trunks of the Scotch firs, the silver birches are aflame with the brilliant yellow of their fading leaves; everywhere the brightness of the colouring makes the landscape joyous.

'In spring and early summer the eye of the traveller is

delighted with a wealth of flowers instead of berries. The Alpine azalea lights up the ground with its profuse display of rosy blossoms, and the purplish-white flowers of the *Anemone vernalis*, and the exquisite pale cream-coloured lily of the valley (*Smilacina bifolia*) give the floor of the valley an aspect of wind-driven foam. The flora of Alpine heights and lowland plains are found flourishing together in vigorous profusion, wild pansies, blue-bells, and ragged-robins being interspersed amongst the bright pink of the *Silene acaulis*, the snowy spikes of *Saxifraga cotyledon*, and the heavenly blue of the Alpine *Veronica*. In Norway the lovers of flowers, mosses, and lichens will find a grand field for study. On the Dovre Fjeld range alone botanists have obtained specimens of as many as 200 mosses, 150 lichens, 50 *Algæ*, and more than 400 phanerogamous plants and ferns.

'The open, park-like scenery of the lightly timbered country charms the eye more than the sombre aisles of the vast fir forests. Eastward of the Dovre Fjeld, where the undulating country slopes away from the Sneehætten and its sister snow-peaks, the traveller on his way north to Trondhjem skirts for many miles a mighty forest of spruce and Scotch fir. From the little posting station of Garlid, perched on a sunny spot above a brawling trout stream, the eastward view is singularly impressive. An immense tract of country is embraced in the landscape, and away to the blue horizon nothing but trees and sky can be seen, the swelling hills everywhere clad with a dense growth of timber and presenting a strangely stern and dusky aspect even in the brightest sunshine. Cross the river, strike boldly into the heart of the forest, and you will find that its appearance from afar is belied by a close insight of its dim recesses. Here propinquity, not distance, lends enchantment to the view. The huge trunks of the firs are not so thickly grouped that the sunbeams cannot chequer with bright patches the needle-sprinkled ground. Here are broad tracts of the beautiful stag's-horn moss and the bright *moltebære* plant, clumps of *woodsia* and the delicate oak fern; the rich fragrance of the pink flowers of the *Linnaea borealis* mingles with the resinous odour of the firs; masses of lichens clothe the rocks; the graceful foliage of birch and aspen diversify the scene; and the stately spires of the spruces tower above all; whilst from out the silent forest, between the ruddy trunks, the eye wanders far away westward

> " To distant mountains, where a thousand peaks
> Flush to the crimson of the dawn's first beam
> Or sparkle silver splendours to the moon." '

We have no lack of sketches of Norwegian scenery from the pens of British tourists, by whom the country is generally described as a land, the most characteristic features of which are forests and fiords; 'forests whose vastness, and shade, and solitude, and silence banish in an instant all associations with songs of birds, and gay scenery;' and combined with these are 'lakes whose deep seclusion puts to flight images of mere grace and beauty—valleys, which from their depth and gloom one might fancy to be the avenue to abodes of mere mysterious creation; mountains, whose dim, and rugged, and gigantic forms seem like the images of a world which one might dream of but never behold.' Thus is the country spoken of by one intrepid traveller, writing under the *nom-de-plume* of Derwent Conway; and in similar terms is it spoken of by others.

Interspersed in many of the volumes referred to are beautiful little sketches of woodland scenery and of the woodland population. Derwent Conway's tour was made, if I mistake not, in 1827. The following sketches of woodlands and wilds I cite from a volume entitled *Rambles in Norway in* 1848 *and* 1849, by Thomas Forester, Esq., as being true still of many such places in Norway, notwithstanding the lapse of time which has occurred since the sketches were penned.

'To regain the valley of the Nid,' he tells in one place,

'we had to cross a tract of country of the wildest character. It was for the most part densely covered with the primæval forest. In many places the tall spruce towered to the height of from 100 (as I calculated) to 150 feet, and were of unusual girth; and the great bulk of the giants of a former generation, which lay mouldering in slow decay, told that no hand of man had been there, as in districts more accessible, to appropriate the statliest of the products of the wilderness. Nature reigned in all her solitary majesty: her operations were uncontrolled. Every age was there; from those lofty piles standing erect in the ripe fulness of their majestic forms, to the young growth that, springing up in every clearance over which the tempest had swept, told of their direct descent from the patriarch of a hundred years, whose crumbling ruins they shrouded with a graceful shrubbery. We count the races of man: who shall say how many generations have here successively germinated and sprung up in youthful vigour and beauty, in a maturer age have hung out from their feathered boughs those pendent tassels of cones, the seed pods of which were destined to perpetuate their species—have ripened, decayed, and gone to dust—since the epoch of the great catastrophe which moulded these wild regions into their present form, and left their bared surface to the gentle and uniform operations with which vegetation—following in the track of ruin—effaces its hardest features, and renews the face of the earth? Touching images from the earliest times have been drawn from the fall of the leaf, as in successive years the short-lived progeny of a single season are thrown of from their parent stems. How much more striking the contemplation of the processes of nature in growth, decay, and reproduction, on the scale on which it is presented in the depths of a primæval forest!

'The general character of the country was irregular, with no leading valleys and few levels of any extent. We mounted ridges of the steepest declivity, where the stunted pines told of the elevation at which we had arrived, to plunge on the other side far down into the depths of dark ravines, through which poured impetuous torrents, chaffing against the smooth cliffs through which they had worn their channels, and eddying round the detached masses which obstructed their course. Clearings, signs of cultivation, and habitations of men, were, as may be supposed, of rare occurrence. The tract we pursued could hardly be called a road; but trains of light, yet hardy, horses, heavy laden with packs, scrambling up the passes, or browsing on the rough herbage, while the rude drivers were seated under the pines, smoking their short pipes, or taking their repasts from the huge leathern pouches which form the invariable equipment of the Norwegian, whether travelling by land, or embarked on lakes or fjords. These groups indicated that such roads were the only means of communication between the towns and ports on the seaboard.' But he shows that even there roads indicative of great engineering skill and power were being created. And having done so, he goes on to say :—

'The nature of the country precluded our having extended views, and only as we were rapidly descending towards a lake which presented in the foreground a wider sheet of water than any we had yet seen, all glowing in the noontide sun, we caught sight of a fine range of hills, I must not call them mountains, though their elevation was considerable, and the outline bold and clearly defined, stretching away to the north-west at a distance of some thirty or forty miles. Just before we had met with the first bed of lilies of the valley in their native habitat, nestling in the shade of an alder copse; and not a furlong beyond we fell in with a herd of these delicate-looking cows, diminutive in size, almost deer-shaped, dun-coloured, and docile in their habits, with which we afterwards became familiar, and which form the staple of the wealth of those pastoral districts towards which we were making progress. Every new object was hailed with fresh bursts of delight.'

In another connection he gives the following account of the forest on the plains —

C

'Crossing the cultivated grounds we immediately entered a forest, the features of which were of an entirely different character from those we had passed in the earlier part of the day. The surface was nearly level for the whole space we traversed that evening and the early stage of the morrow—a distance of eighteen or twenty miles. It lay along the left bank of the Nid, which on its other shore washed the base of that long range of perpendicular cliffs which we had marked from our last station. There was no undergrowth, except where we occasionally crossed water-courses, which discharged themselves into the river. The banks of these were profusely hung with alder and birch. The boles of the tall pines were also clear of boughs to the height of fifty or sixty feet. Upwards, their tapering stems and spreading branches were of a bright resinous hue, to which the rays of the setting sun gave additional lustre, in singular contrast with the hoary cast of the scaly trunks below, to which the shades of evening already imparted a deeper tint. The trees appeared as regularly set out as if they had been artificially planted and thinned—one looked in vain for those giants of the forest which had before attracted our notice. No prostrate masses, moulding in gradual decay, told the tale which had before led us to moralise on the processes of nature and the revolutions of time. The rocky steeps, the rough and tangled brake, all which before had given that air of savage wildness to the forest, were here wanting. But still the sandy plains which we were now traversing had a character of magnificence peculiarly their own. The wider extent of the same unbroken line canopied above by that dark mass of spreading foliage; those countless columns, which, far as the eye could reach in every direction, mile after mile, stood tall, erect, dignified —supporting that living roof; these long drawn vistas, through the receding arches of which one sought in vain to penetrate the depths of that vast solitude; the deepening gloom still chequered by the rays which the setting sun shot athwart the trees; the silence unbroken save by the roar of the river, our constant, though for the most part unseen companion, as it hurried down the frequent rapids;—all this gave a new and solemn phase to our thoughts.'

But there is yet another picture needed to illustrate fully the forest scenery of Norway; and this the volume which I have cited supplies. The author subsequently writes:—' Before we take leave of the valley of the Nid, I must endeavour to give a brief sketch of its most striking scenes. On the skirts of the forest we again struck the river, flowing silently, deep and glassy, with a strong current to the southward. But we could just perceive, and our ears received distinct intimation, that its character was about to change. Having crossed a ferry just below some rapids, over which it was tumbling in angry confusion, the wild roar of the waters increased. About a mile above, the whole body of the river is projected over a ledge of rocks, forty or fifty yards in width, which dams up the breadth of the channel. As yet, however, only the upper edge of the fall was visible. Seen at a distance, above a screen of firs, the long white sheets of foaming water, stretching from bank to bank, appeared like folds of linen extended on the racks of a bleaching ground. There are three successive falls, of which the principal and most precipitous, where the river, confined in deep clefts, turns a sharp angle, may not exceed fifty or sixty feet in height. But though that is insignificant compared with many others which we afterwards saw, the depth of the fall itself is not the only ingredient in the grandeur of such a scene. The broad sheet, and comparatively small elevation of this, put me somewhat in mind of the falls of Schaffhausen. But its most remarkable feature was the immense quantity of timber which, having floated from the upper country, was here carried down the current. The enormous logs, first whirled, fearfully booming, against the rocks that narrowed the channel, were then hurled over, and plunged into the boiling foam below, At the foot of each fall a perfect barrier of pines was formed, to which many were added while we stood witnessing the struggle. Some, eddying in the whirlpools, seemed destined never to get free; one almost wondered how any escaped: numbers were broken up, and some never recovered. The whole shore below the falls was strewed with the giant bulk, *disjecta membra*, of these spoils of the forest, thus arrested in their progress to the sea.

'Felled and sledged to the nearest stream during the winter, no sooner is its frozen channel set free by the returning spring, and swelled by the influx from the dissolving snow, than the timber, thus left to its fate, begins its long journey Borne down by the foaming torrents which lash the base of its native hills, far in the interior; hurried over rapids; taking in its onward course along the shores of winding lakes, or slowly dropping down in the quiet current of broad rivers, the accumulated mass is brought up at last by a strong boom placed across the stream, where it discharges itself into navigable waters. It is then sorted, appropriated by the merchants to whom it is consigned, and shipped for foreign ports. One would wonder how it ever reached the place of its destination, or how, of the numerous owners, each could recognise his own. But I was given to understand that the logs are branded with the owner's mark before they are committed to the stream; and I observed that during their passage down the lakes they were collected into immense rafts, curiously framed and pinned together; but so unwieldy and unmanageable are the masses, that but little can be done in the way of navigation beyond fending them off the shores and rocks, and keeping them in the current. Some of the timber is said to be two years in finding its way to the coast.'

In accordance with this narrative is that already given of what was seen at Vigelund, on the Torrisdal river, flowing into the Fiord at Christiansand.

To resume the narrative by Forester:—

'A turn in the road brought us in view of a scene of desolation on a magnificent scale. Fire had done its work of devastation, and, running up the tangled banks of a wild ravine, had penetrated far into the vast recesses of the forest. The jagged and charred stems of the pines, snapped asunder at various heights—the blackened and calcined rocks—the screen of shrivelled spray that hung withered from the half-burnt trees at the line where the conflagration had stayed its devouring course, formed a spectacle among the most striking that can be conceived. The scene of wreck enabled us to form some faint idea of what it must have been when the conflagration was at the height of its fury, passing in its conquering strength from tree to tree, spreading through the tangled branches, climbing in wreaths of flame the tall stems, till it overtopped the highest summits—amidst volumes of smoke and jets of sparks, and the crackling and roaring of the destroying element, and the crash of falling trees, as one after another came to the ground.'

With one other sketch of forest scenery I close. It is one from the vicinity of Hurungerne, and is of Skagastöls-Tind, the highest mountain in Norway, reaching a height of 7670 English feet, situated in the Koldedal, or Cold Valley:—
'It was a snowy walk of ten or twelve miles to Koldedal, into which we slid from the fjeld at a frightful velocity down a snow-drift. Some fine *tinds* [corresponding to the *aiguilles* of Swiss scenery—pointed rocks projecting to a great height] rise over Koldedal. The view from the foot of the first descent was inconceivably grand. The clouds occasionally concealed the whole horizon, and then, breaking, revealed the jagged peaks of Hurungerne rising above them in wild, fantastic confusion, and Skagastöls-Tind. Beneath, a bell-shaped snowy valley penetrated into the mountains, and was closed by a vast glacier. Almost all the points were black and bare, rising like *aiguilles* from the masses of snow which overspread the lower ranges of the fjelds; the summits themselves, though they are from

7000 to 8000 feet high, being too steep to hold snow on their surface, and attracting powerfully the sun's rays, the slight coating of snow is soon melted. We were disposed to linger long in view of this glorious spectacle, and it was with difficulty our guide drew us away. Our path lay down the Koldedal, soon coming among birch woods, and afterwards entering the most splendid fir forest I ever met with. Gigantic trees lay about in all stages of decay; some stood scathed, with naked arms bleaching in the weather; others were vigorous and of enormous growth. One we measured was nearly four feet in diameter. And to increase the grandeur of the forest scene, the peaks of the Hurungerne were seen rising above the pines in the background.'

Such are the woodlands of Norway. By one traveller, Norway is spoken of as a land 'whose only charm is to be found in her dim mountains, her silent forests, and her lonely lakes.'

Another, Edward Price, an artist, who traversed the land and looked upon every scene with an artist's eye, speaks of Norway as a country 'which surpasses every country of Europe in the depths of its fiords, and in the grandeur of its forests and forest scenery.' Having landed at a distant point, and traversed the land, chiefly on foot, seeing thus much which could not otherwise have been seen in the course of his tour, he reached the capital; and of what he saw as he approached it he thus writes:—
'Luxuriant pasturage and crops, giving rich promise of an abundant harvest, lay on every side. Wood was no longer the great staple of the land, but was scattered over a charming undulating country only in such quantity as served to shelter the fields and beautify the landscape; nor was it now confined to fir, but included all the variety of trees which we are accustomed to find in the temperate altitudes. The Christiania Fiord, spotted with its islands, and seemingly environed with its finely wooded banks, formed innumerable bays and creeks, all calm and pellucid beneath the warm rays of the noonday sun.'

CHAPTER III.

MOUNTAIN PLATEAUX AND MOUNTAIN RAVINES.

AMONGST the export timber ports of Norway, Avendal, Drammen, Christiania, and Fredrickshold, an important place is held by Drammen, the harbour of which is always crowded with tiers of shipping, with huge piles of timber on the wharfs, and vast rafts of timber afloat upon its waters. From this port some 110,000 tons of timber are annually shipped to England, Holland, and France.

After reaching Christiania my first excursion, I have said, was towards Drammen. There M. du Challu had been before me; and instead of telling here of what I saw, beyond stating that I greatly admired the trim cottages and well kept fields through which the train passed on leaving the capital, I would take my readers further in the wake of M. du Challu, in the journey which he made through Drammen and Konsberg to the west coast. Of which he has left an account, in which he tells with pleasure of the peculiar costume and habits of the Saetersdal, and of the Lemarken, of which there are two divisions. He tells of a drive of twenty miles from Konsberg, which brought him to a forest on a plateau 1700 feet above the level of the sea, whence descending a ravine through a dark wood, he found suddenly burst into sight the farm of Bolkesko, 1240 feet above the level of the sea, of which he writes:—
'I know of no farm in Norway so picturesquely situated, and none with such peculiarly superb landscape. It was nestled among fir-clad hills, whose dark colour contrasted with the green meadows and fields which they surround. The place was partly hemmed in by barren mountains, on which were patches of snow. Here in a steep valley two lakes apparently overlapping each other are noticed: the

Bolke, of a triangular shape, 1000 feet; and a little beyond this the Tol, 690 feet above the sea level. Everywhere little streams trickled down the hill-side, filling the air with the sweet music of their waters.' The house and its inhabitants he found to be not less interesting than the house was beautifully situated. About seventeen miles west of this farm he came upon the lower end of Tin lake, upon which there plies a little steamer; and he tells:—' The shores of the lake are thoroughly Norwegian, with rugged mountains covered with forests to their very tops. Toward the northern portion of the western shore one enters a part of the lake called Vestfiord, running east and west; the scenery increases in beauty, the landscape reminding one of the Hardanger. Leading from this fiord is a fine narrow valley called Vestfiordal, on the left of which Gaustad rises 6000 feet high. It is celebrated for the Ryukandfoss, at its end, one of Norway's highest and most beautiful waterfalls. The valley terminates abruptly, closed by gigantic walls, but the spray of the turbulent waters is seen long before the fall is reached. The *Ryukandfoss* [reeking or smoking waterfall], plunges into a chasm from a height of 780 feet over a perpendicular ledge on the table-land. It is formed by the river Maan, which rises in the Mjoes Vand. The sight is appalling as the eye seeks the depth below amidst the roar of the waters: it is a fascinating spot.'

Perhaps we too may find it fascinating, so fascinating as to lead us to loiter by the way, and forget the length of the journey which is before us; but I do not suppose any one will find the time mis-spent.

To make more intelligible the account of this region which I have to give, I may premise that in the northern portion of Norway the land presents the appearance of table-lands, or comparatively level plateaux, cut up by what may be called ravines rather than valleys, which are sometimes more than a thousand feet in depth; these can only be crossed by zigzag tracks or roads, descending the precipitous declivity on one side, crossing a streamlet at the bottom, and ascending in a similar zigzag way a corresponding precipitous ascent on the other; and in places, that of isolated hills and mountains, scattered about in what looks like studied confusion, sometimes standing apart and alone, but as frequently in groups of more or less irregularity, and of greater or less extent; and sometimes, but that rarely, taking a form not unlike a mountain range. Towards the south the country assumes gradually a more level aspect, but it does so without losing altogether its hilly character. The result of the whole is that about two-thirds of the country is at an elevation of upwards of 2,000 feet above the level of the sea, which is considerably above the range of forest trees in that land. And so far south as this, *plateaux*, *fjelds*, or fields of high altitudes cut up by deep ravines may be found.

Of the ravines by which the blocks of the high-lying table land are separated, a definite and probably correct conception may have been formed from what has been stated. They, and the fjelds with which they are connected, are characteristic of the region in which they exist. Most, and perhaps all of them, may be crossed in the way spoken of, by zig-zag descending and ascending footpaths, at some place or another; but many, if not most of them, are at other points impassable; and of the appearance presented by them throughout a great part of their extent an illustration is supplied by what is known in this locality as the *Marie Stege*, or Mary's Ladder, near to this Ryukand Foss. Marie, whose name is given to the precipice, lived on a farm situated on the side of a mountain, which there blocks up the valley. According to the facts of the case, in the local story, as told by Williams : 'The ordinary track by which the lower part of the valley may be reached ascends about a thousand feet over the ridge of this mountain, and then, of course, a corresponding descent has to be made. But the river—how does that find its way down the valley ? There must be some way for it. Yes; there is a deep clift, a great chasm, more than a thousand feet (some say two thousand) feet in depth. Ages had passed away, and nobody had dreamed of any other way to reach the lower valley than that over the mountain; but Marie, whose lover lived below, had heard of a plot by his rival to waylay him as he came by a track over the ridge to visit her, so she tried the dreadful precipice, and found that by clinging with fingers and toes to the little ledges of the rock she could pass in a direct line along the face of it. She thus warned her lover of his danger, and enabled him to meet her secretly and safely by traversing the giddy path she had discovered; and the lovers evaded—as lovers always do—both the cruel father and his accomplice, the wealthy rival. By this path they met as usual, until at last detected; and then Ejotein Halfoordsen, the lover, was prevailed upon to fly in order to escape new plots upon his life. In the course of years the father died, the rival ceased to persecute, and Ejotein returned with fame and wealth. He came by the shortest way. Marie saw him coming, and called his name aloud; he raised his arms and waved his hands as a signal of recognition, and by doing so was overbalanced and fell. She watched his falling body till it disappeared in the foam of the Rinkan Foss; when the dark veil of madness fell over her mind, and fulfilled its beneficent intent by shutting out a knowledge too horrible for endurance!'

Williams having come to the farm, got a little girl to guide them to the edge of the precipice, whence a distant view of the Foss is obtained, and to show him the beginning of the track leading to the descent to the Stege itself—which, however, she had been forbidden to go near. 'I then,' says Williams, 'proceeded along what I supposed to be the Marie Stege, a ledge of rock trodden with footslips, varying from six inches to a foot in width, with a sloping wall of rock above and the chasm below; this continued until I came to a part where there are two tracks, one apparently leading over the hill, the other direct to the perpendicular wall of the precipice, which is seen a little further on, rising to a fearful height overhead, and proceeding downwards to the gulf below, with an unbroken smoothness that looks utterly hopeless : but I determined to go on as long as there was any vestige of a track. Following thus the marks of footsteps, I came out at last upon the edge, but upon the face of the precipice, which is formed by the splitting down of the barrier mountain before referred to ; it was a giddy path, but I kept along it, placing my feet upon the worn ledges, and clinging to others above until I came to a tree which grew upon a ledge similar to those I had stepped upon, but much wider, and which seemed to be the end of the track I was following. Some initials cut upon the tree, as triumphant indications of the carvers' exploit in reaching it, rather confirmed the notion that I had only followed a track leading to this as a station for viewing the waterfall and the whole of the great chasm, which are well displayed from this point.

'Concluding that such was the case, and that the other ascending track leads to the Marie Stege, I was about to return, when I saw, far away below me, standing on a large table of rock, five student-looking young men, with a peasant, who appeared to be their guide : they hailed me, and I returned their salutation, but could not hear what they said. Then the peasant took off his shoes, left them, and presently reappeared, moving along the face of the precipice like a fly upon a wall. His means of adhesion were totally unintelligible from the distance, but as he approached I perceived that he was clinging by fingers and toes to narrow ledges of rock, from one to four or five inches in width. At last he reached me, and asked me whether I would accompany him back, which I consented to do, though it appeared rather a dangerous exploit. I found, however, that it was much easier than it appeared to be from the distance. The rock has a perpendicular lamination, and doubtless a cleavage to which the formation of the chasm is due; the abrupt termination of these laminæ form ledges, which, though very narrow, are perfectly firm and safe, affording a reliable foothold without the slightest tendency to slippiness; besides these there is an abundance of similar ledges, affording firm fingerhold, which, though but an inch wide, give a most comfortable assurance of safety to the climber, who, bending the hands claw fashion, clung to them with the finger ends. I would rather, under such circumstances, have a firm two inch foot ledge, and one inch of such finger hold, than an eighteen inch pathway with nothing for the hands. At about half-way I stopped to contemplate the scene, which is magnificent, and its grandeur is heightened by the peculiar position from which it is seen.

'Imagine yourself "holding on by your eyelids," as the sailors have it, in the manner just described, to the face of a precipice which rises overhead some five hundred or six hundred feet, the upper part being, in fact, quite out of sight; then, with great care, and some fear and trembling, you turn yourself round, gradually placing your heels on the former position of your toes, removing your hands, one at a time, from your clutching place, and finding a lower ledge upon which to rest the wrist-end or heel of the hand. Having anchored yourself thus, and keeping your back quite flat against the rock (as any leaning forward would be fatal), you look in the direction of the upper part of the valley, and see far below, and far away, a dark chasm, partly hidden by branches of trees; through this the river flows, and as it comes nearer reaches a wider opening of the gorge, advancing towards the edge of a precipice, over which it rolls to a gully of its own cutting, and then pitches down an unknown depth, for a white cloud hides the bottom of the dark abyss, and rises high into the sunshine. This is the perpetual spray—the reeking, or "rinken," from which the name of the fall is derived. You may, however, estimate the depth of the fall, for, looking down the grey wall to which you are clinging, you see that its gully terminates in dark, quiet water. This is the same water that a few minutes ago was tearing down and thundering so furiously, and partly rising again to form the ever hanging, though ever falling cloud.'

On arriving at the end of the Marie Stege, and landing upon the platform of rock where the five tourists stood, Williams was congratulated on what they considered his narrow escape: they being under the impression that when they saw him turn round to look at the waterfall he had reached a point from which he could neither advance nor recede. And they were surprised when they learned the truth—nor is this to be wondered at, as viewed from their position, where the ledges are quite invisible, and both the height above and the depth below are fairly seen, it must be a somewhat thrilling sight to witness the crossing of the Marie Stege—far more so than to do it.

The broad platform of rock affords by much the best view of the fall, and those who come from below have no occasion to cross the Marie Stege, except for its own sake.

But where is Du Challu? He is on and away, and by this time half way up the beautiful Fiord of Hardanger. We cannot overtake him; but we may encounter him again on the Sogne Fiord further to the north, situated between this and the forests of Romsdal county, of which mention has been made. Seeing that our leader has gone on let us take another look at the fall. This waterfall is frequently visited by tourists, but it is generally, as mentioned by Williams, approached from below. Bayard Taylor who did so gives the following description of the scene in his volume entitled *Northern Travel.* Writing of his journey towards the Tindso, he says:—' During this stage of twelve or thirteen miles the quality of our carioles was tested in the most satisfactory manner. Uphill and down, over stock and stone, jolted on rock, and wrenched in gulley, they were whirled at a smashing rate, but tough ash and firmly-welded iron resisted every shock.

For any other than Norwegian horses and wheels it would have been hazardous travelling.* It was dark when the new horses came; and ten miles of forest lay before us. We were ferried one by one across the Tind Elv on a weak, loose raft, and got our carrioles up a frightful bank on the opposite side by miraculous luck. Fortunately, we struck the post-road from Hetterdal at this place, for it would have been impossible to ride over such rocky bye-ways as we had left behind us. A white streak was all that was visible in the gloom of the forest. We kept in the middle of it, not knowing whether the road went up, down, or on a level, until we had gone over it. At last, however, the forest came to an end, and we saw Tind Lake lying still and black in the starlight.

'In the morning we took a boat with four oarsmen for Mael, and the mouth of the Westfiord-dal, on which lies the Rinkan Foss. There was no end to our wonderful weather. In rainy Norway the sky for once had forgotten its clouds. One after another dawned the bright Egyptian days, followed by nights soft, starry, and dewless. The wooded shores of the Long Tind Lake were illuminated with perfect sunshine, and its mirror of translucent beryl broke into light waves under the northern breeze. . . . The highest peaks rise to the height of 2000 feet, but there is nothing bold and decided in their forms, and after the splendid fiords of the western coast the scenery appears tame and commonplace. The boatmen pulled well, and they slept at Ole's *gaard*, situated on a turfy slope, surrounded with groves, above the pretty little church of Däl, halfway between Mael and the cataract. They arrived here about four o'clock, when the sun was about resting his chin on the shoulder of the Gousta; and they must visit the fall and return.

---

* Such excursions must be made in *Carrioles*, the usual conveyance of the country They may be described as very low seated long gigs, capable of seating only a single person, who sits in a half-reclining posture. This is placed on very long shafts, which are more or less elastic, and, extending behind for a considerable length, supply a resting place for luggage, and for the lad who may have to fetch back the conveyance or the horse to the station whence it may have been hired.—J. C. B.

'Having refreshed themselves with a bottle of Bavarian beer, and ordered supper and bed, they pressed on. There were still ten miles to the Rinkan, and consequently no time to be lost. The valley contracted, squeezed the Maan between the interlocking bases of the mountains, through which, in the course of uncounted centuries, it had worn itself a deep groove, cut straight and clean into the heart of the rock. The loud perpetual roar of the vexed waters filled the glen; the only sound except the bleating of goats clinging to the steep pastures above us. The mountain walls on either hand were now so high and precipitous that the bed of the valley lay wholly in shadow; and on looking back its further foldings were dimly seen through the purple mists; only the peak of the Gousta, which from this point appeared entire and perfect.

'The valley of the Maan, apparently a rich and populous region, is in reality rather the reverse. In relation to its beauty, however, there can be no two opinions. Deeply sunken between the Gousta and another bold spur of the Hardanger, its golden harvest fields, and groves of birch, ash, and pine, seem doubly charming from the contrast of the savage steeps overhanging them, at first scantily feathered with fir-trees, and scarred with the tracks of cataracts and slides, then streaked only with patches of grey moss, and at last bleak and sublimely bare. The deeply channelled cone of the Gousta, with its indented summit, rose far above us, sharp and clear in the thin ether; but its base, wrapped in forests, and wet by many a waterfall, sank into the bed of blue vapour which filled the valley.

'Noon brought us to Hakenaes, a distance of twenty-one miles. Here we stopped to engage horses to the Rinkan Foss, as there is no post station at Mael.' They were informed that the horses would be at Mael as soon as they, and they resumed their seats in the boat.

They arrived first, and lay upon the bank for some time after arriving there, watching the postillions swim the horses across the mouth of the Maan Elv. 'Leaving the boat which was to await,' says he, 'our return the next day, we set off up the West Fjord-dal towards the broad cone-like mass of the Gousta Fjeld, whose huge bulk, 6000 feet in height, loomed grandly over the valley. The houses of Mael clustered about its little church, were scattered over the slope above the lake; and across the river, amid fields of grass and grain, stood another village of equal size. The bed of the valley, dotted with farms and groups of farm houses, appeared to be thickly populated; but, as a farmer's residence rarely consists of less than six buildings—sometimes eight—a stranger would naturally overrate the number of its inhabitants. The production of grain also is much less than would be supposed from the amount of land under cultivation, owing to the heads being so light.

'The pyramid, 1500 feet in perpendicular height above the mountain platform from which it rose, gleamed with a rich bronze lustre in the setting sun. The valley was now a more ascending gorge, along the sides of which our road climbed. Before us extended a slanting shelf, thrust out from the mountain, and affording room for a few cottages and fields; but all else was naked rock and ragged pine. . . .

'When we reached the little hamlet on the shelf of the mountain, the last rays of the sun were playing on the summits above. We had mounted about 2000 feet since leaving Tind Lake, and the dusky valley yawned far beneath us, its termination invisible, as if leading downward into a lower world. Many hundreds of feet below the edge of the wild little platform on which we stood thundered the Maan in a clift, the bottom of which the sun has never beheld. Beyond this the path was impracticable for horses; we walked, climbed, or scrambled along the side of the dizzy steep, where, in many places, a false step would have sent us to the brink of gulfs whose mysteries we had no desire to explore. After we had advanced nearly two miles in this manner, ascending rapidly all the time, a hollow reverberation, and a glimpse of profounder abysses ahead, revealed the neighbourhood of the Rinkan. All at once patches of lurid gloom appeared through the openings of the birch thicket we were threading, and we came abruptly upon the brink of the great chasm into which the river falls.

'The Rinkan lay before us, a miracle of spraying splendour, an apparition of unearthly loveliness, set in a framework of darkness and terror befitting the jaws of hell! Before us, so high against the sky as to shut out the colours of sunset, rose the top of the valley, the level of the Hardanger table-land, on which a short distance further lies the Mios Vand, a lovely lake in which the Maan Elv is born. The river first comes into sight a mass of boiling foam, shooting around the corner of a line of black cliffs which are rent for its passages, curves to the right as it descends, and then drops on a single fall of 500 feet into a hollow cauldron of bare black rock. The water is already foam as it leaps from the summit; and the successive waves, as they are whirled into the air, and feel the gusts which for ever revolve around the abyss, drop into beaded fringes in falling, and go fluttering down like scarfs of the richest lace. It is not water but the spirit of water. The bottom is lost in a shifting snowy film, with starry rays of foam radiating from the heart, below which, as the cloud shifts, break momentary gleams of perfect emerald light. What fairy towers of some Northern Undine are suggested in those sudden flashes of silver and green! In that dim profound, which human eye can but partially explore, in which human foot shall never be set, what secret wonders may still lie hidden! And around this vision of perfect loveliness rise the awful walls wet with spray which never dries, and crossed by dazzling turf from the gulf below our feet, until still further above our heads, they lift their irregular corners against the sky.

'I do not think I am extravagant when I say that the Rinkan Foss is the most beautiful cataract in the world.

I looked upon it with involuntary suspension of the breath and quickening of pulse, which is the surest recognition of beauty. The whole scene, with its breadth and grandeur of form, or its superb gloom of colouring, which enshrines this one glorious flash of grace, and brightness, and loveliness, is indelibly impressed upon my mind. Not alone during that half hour of fading sunset, but day after day, and night after night, the embroidered spray-wreaths of the Rinkan were falling before me.

'We turned away reluctantly at last, when the emerald pavement of Undine's palace was no longer visible through the shooting meteors of silver foam. The depths of West-fjord-dal were filled with purple darkness; and the perfect pyramid of the Gousta, lifted upon a mountain basement more than 4000 feet in height, shone like a colossal wedge of fire against the violet sky. By the time we reached our horses we discovered that we were hungry, and, leaving the attendants to follow at their leisure, we urged the tired animals down the rocky road. The smell of fresh-cut grain and sweet mountain hay filled the cool evening air; darkness crept under the birches and pines, and we no longer met the home-going harvesters.'

It was between nine and ten ere they reached Däl; and in the dark they disturbed the sleepers in more than one *gaard* before they reached their resting place. There every arrangement was made for their comfort. There was a white cloth on the table in the guest's house, some roast mutton, potatoes, and beer, in praise of all which the traveller waxes eloquent, graced the board, but there was only emptiness when they arose—they had consumed all. In the upper room there were beds with linen fresh as youth, and aromatic as spring; 'and,' says he, 'the peace of a full stomach and clear conscience descended upon our sleep.

'In the morning we prepared for an early return to Mael, as the boatmen were anxious to get back to their barley fields. I found but one expression in the guest's book—that of satisfaction with Ole Tergensen, and cheerfully added our amen to the previous declarations. . . .

'We bade farewell to the good old man, and rode down the valley of the Maan, through the morning shadow of the Gousta. Our boat was in readiness; and its couch of fir boughs in the stern became a pleasant divan of indolence after our hard horses and rough roads. We reached Tinoset by one o'clock, but were obliged to wait until four for horses. The only refreshment we got was oaten bread and weak spruce beer. Off at last, we took the post road to Hitterdal, a smooth excellent highway, through interminable forests of fir and pine. Towards the close of the stage glimpses of a broad, beautiful, and thickly settled valley glimmered through the woods, and we found ourselves on the edge of a tremendous gully, apparently the bed of an extinct river. The banks on both sides were composed entirely of gravel and huge rounded pebbles the masses of which we loosened at the top, and sent down the sides, gathering as they rolled, until, in a cloud of dust, they crashed with a sound like thunder upon the loose shingles at the bottom, 200 feet below. It was scarcely possible to account for this phenomenon by the action of spring torrents from the melted snow. The immense banks of gravel which we found to extend for a considerable distance along the northern side of the valley seemed rather to be a deposit of an ocean flood.

'Hitterdal, with its enclosed fields, its harvests, and groups of picturesque farm-houses, gave us promise of good quarters for the night; and when our postillions stopped at the door of a prosperous-looking establishment we congratulated ourselves on our luck. But—Never whistle until you are out of the woods!' They met with sorry welcome, sorry entertainment, and sorry fare. 'We did not ask for coffee in the morning, but as soon as we could procure horses, drove away hungry and disgusted from Bamble-Kaasa and its respectable inhabitants. We passed the beautiful falls of the Tind Elv, drove for more than twenty miles over wild piny hills, and then descended to Kongsberg, where Fru Hansen comforted us with a good dinner. The next day we breakfasted in Drammen, and in baking heat, and stifling dust, traversed the civilised country between that city and Christiania.'

The peculiar feature of the physical geography of the country, thus brought under consideration, gives character to the exploitation of the forests, in connection with the bringing out of the timber, as will afterwards be detailed.

CHAPTER IV.

GEOGRAPHICAL DISTRIBUTION OF TREES IN NORWAY.

IN 1876 there was published a valuable report on *The Kingdom of Norway and its People*, by Dr. O. J. Broch, a distinguished statistician of European reputation, and President of the Norwegian Commission for the International Exhibition in Paris in 1878, in which was embodied a sketch of the vegetable products of Norway, including forest trees. And in 1879 there was published as a contribution to the jubilee connected with the 400th anniversary of the founding of the University of Copenhagen, *A Report on the Distribution of Plants in Norway*, by Dr. F. C. Schuebeler, Professor of Botany in the University of Christiania, in which is supplied much additional information in regard to forest trees, indigenous or cultivated, in Norway, with plates illustrative of the appearance of these and of the form and venation of their leaves. There are given with this report, charts illustrative of the mountainous region of the country; of the coast lands of northern Norway; of the temperature throughout Europe on 1st January, and in the month of July; of the lines of minimum temperature throughout north-eastern Europe; of rainfall; of the humidity; and of the geological strata superimposed on the primitive rocks. And in the report is embodied the longitude and latitude of the most northern localities in Norway in which have been observed the indigenous growth, or cultivation, or flowering, or fruiting—it being specified which it is—of well-nigh 4000 different plants found there. Along with this there was issued a large map of the kingdom of Norway in four sheets, illustrative of the same, together with a corresponding map in one sheet, with coloured indications of the forests of coniferous and of broad-leaved trees in the kingdom. And subsequently there was issued by the Directory of Forest Administration in Christiania three forest maps of different districts of the county of Christiania. From these it appears that broad-leaved trees are found chiefly in the west and the north-west portion of the kingdom, though by no means confined to these, and mostly on the borders of rivers and lakes, and skirting the lower fringes of coniferous forests; while the forests of coniferous trees are densely diffused over the south-east, and sparsely scattered over the extreme north of the kingdom.

Besides these I have before me a Report by the Forest Directory for the period 1875-1880; a Report on the Nature and Condition of Forests in Finmark, by Forest-Inspector Barth; a Report on the Nature and Condition of the Forests in the Guldbrandsdalen, by the same author; a Report on the Condition of the Woods in the Watercourses of the Arendal, by Forest-Inspector Mejdell; a Report on the Condition of the Forests in Romsdals County, by J. Schioetz; and Reports by A. T. Gloersin; and of Examination of Forests in Stavanger and the borders of Bergenuus County. Also reports on the Economical Condition of the Kingdom by Prefects of nineteen Prefectures, for the years 1861-65; similar reports for the years 1866-70; and a Report made by Forest-Assistant Aars, to the Department of the Interior, on the Condition of the Forests in the Prefecture of Lister and Mandal, published in successive numbers of the Christiansand *Stiftsavis*, in the latter months of 1870, containing saddening accounts of the reckless and hopeless destructions which were then, and had been for some time, going on in the forests.

From the whole it appears that the true forests of Norway are composed almost entirely of the Norway spruce fir and the Scots fir. It is only exceptionally that some other trees, such as the alder, the beech, and the oak, though not numerous, are found forming little woods.

Of the pine, *Pinus sylvestris*, Dr Broch reports that it is found everywhere throughout the country; in the south to an elevation of 950 and 900 metres; on the Nordre-Guldbrandstal, 62° N. lat., to 900 and 800 metres; in the Diocese of Drontheim, 63° to 65° N., to 650 and 500 metres; in the prefecture of Nordland to 550 and 350 metres; but in Finmark, 70° N. rarely at more than 200 metres above the level of the sea.

The Norway spruce fir is generally known as *Abies communis*. According to Hullet, this generic name is derived from, and is of the dialect of, the Celtic *Abetoa*, whence come *Abiete* Italian, *Abeto* Spanish,&c. Hesychus, the Greek grammarian, calls it *Abin*. According to others, the name is derived from the Latin *Abeo* to spring—and has been given in reference to its lofty and aspiring habit—or the Greek *Apios*, a pear tree or a pear, the name being given to this tree in reference to the form of its fruit! Amidst etymological derivations so varied I have no choice. By some it is alleged to be identical with the *Abies excelsa* of De Candolle. It is known as the *Fichtenbaum* of Germany, the *Abiete* of Italy. I have found it spoken of as identical with the *Sapin* of France; but the *Sapin communis*, the silver fir of England, known by several other names in different districts, is the *Picea pectinata*, D. Don, and the *Abies pectinata* of De Candolle; while to the *Abies excelsa* of De Candolle, the *Abies Picea* of Millaw, the *Picea excelsa* of Lank, is given the name of *Sapin rouge*, *Sapin gentil*, *Sapin épicéa*, &c., and now generally the name *Epicéa commune*.

Some confusion, says Loudon, exists in the works of modern authors respecting the silver fir and the spruce, partly, as it would appear, from the circumstance of Linnæus having made an erroneous application of the names given to these trees by Pliny. The tree which

Theophrastus calls *Aale*, Pliny calls *Abies*, and Linnæus *Pinus picea*, while the tree that Pliny calls *Picea*, which is our spruce fir, is named by Linnæus *Pinus abies*. 'This tree is found principally,' writes Dr. Broch, 'in the eastern portion of the country and in the diocese of Drontheim. On the west coast it is now met with growing wild to the south of 62° N., but it is found in plantations in many other places. It is only in the interior of Hordaland at Mo, an annex of Hosanger, 60° 48' N.; and at Voss 66₀ 38' N., that it is met with in a wild state. To the north of Cape Stat it is found along with the pine even on the islands along the coast to 65° N. lat.; beyond that it becomes more rare; and it ceases to form forests somewhere near the Arctic Circle. In east Finmark, on the contrary, it is found in Syd-Varanger, near the lake Kjolmejare, 69° 30' N. lat.; but in separate trees or in quite small groups. In certain localities in southern Norway the fir attains on mountains almost the same elevation as the pine, but, as a general rule, its limitation of vegetation is from 60 to 100 metres below the limit of that tree.

'In Southern Norway the pine and the fir attain to their greatest dimensions, but continually the trees of great size are becoming every year more rare, in consequence of the felling of the large trees increasing with the improvement of roads, and of means of transport. In illustration of the magnitude attained, there was felled some years ago at Lom, in Nordre-Gulbrandsdalen, 61° 53' N., at an elevation of 560 metres, a pine, the trunk of which near the ground had a diameter of 1·2 metre, 4 feet, and a diameter of 0·5 metre, 20 inches, at a height of 16 metres, or 53 feet 4 inches. At Nordre-Aurdal, and Valders, 60° 57' N., there stand still two pines called the *Soesterfuruer*, or Sister Pines, of which the larger, measured in 1864, had a height of 28 metres, 93 feet 4 inches, and a diameter of 1 metre, 40 inches, at the height of 1 metre from the ground. In Northern Norway, and especially in Finmark, these trees have less height, but they present a considerable girth.

'In Drangedal, in the bailiwick of Bamble, 59° 6' N., there has been measured a fir 31 metres, 103 feet in height, and 0·76 metre, 66 inches, in diameter, at 21 metres, 70 feet, from the ground. In 1864, there was felled in Hurdal, 60° 24' N., at an altitude of 350 metres, a fir having a height of 34 metres, 103 feet, and at 1 metre from the ground a diameter of 1·07 metre, 47 inches. There may be counted in it 170 concentric rings; and it was estimated that it must have been from 175 to 180 years old. In Selbo, 63° 15' N., there was felled in 1877 a fir having a height of 33 metres, and a diameter of 1 metre, 40 inches, at 1 metre from the ground.

'The time required by the fir and the pine to attain the dimensions of building timber, and timber for export, naturally varies much, according to the conditions of growth in which they find themselves. In Southern Norway it may be reckoned that a pine of from 100 to 150 years' growth may furnish trees of 5 metres or 17 feet, and of 33 centimetres or 13 inches diameter at the small end. In the forests of Vinger and Odal, 60° to 61° N., in the valley of the Glommen, their concentric rings measured 1 centimetre, nearly half-inch; in the northern part of the valley of Osterdal, on the other hand, it requires from 4 to 6 to measure 1 centimetre; the rings of the fir being less compressed from 2 to 4 often measure 1 centimetre. In young trees the augmentation of girth as elsewhere diminishes continuously with age.'

In a paper transmitted under date of September 21, 1874, to the Department of State at Washington, General Andrews, Minister from the United States to Sweden, draws attention to the fact that a thing may be reckoned as worth what under favourable circumstances, it takes to produce it, and reports:—Mr. Samson, a highly intelligent Norwegian gentleman, who has made a large fortune in the timber trade, informed me some time ago that, according to a calculation he had made, pine and spruce timber actually costs, and is worth much more than the price at which it is sold. His theory is, that an acre of ground timber is worth the same that the lowest or nominal price of wild land—say 1 dol. an acre—would amount to, as an invested capital, drawing interest at the expiration of the period required for timber to develop. 'In the report on Swedish forest culture, accompanying my letter No. 166,' says he, 'it was shown that on the northerly part of Sweden, 200 years, and in poorer soils 300 years, are required for the pine to grow to good timber. In the south part of the country 100 years are sufficient.' He says that 1 dol. invested at 5 per cent. interest per annum will double in 20 years. In 40 years it will be 4 dols.; in 60 years, 8 dols.; in 80 years, 16 dols.; in 100 years, 32 dols.; in 120 years, 64 dols.; in 140 years, 128 dols.; in 160 years, 256 dols.; in 180 years, 512 dols.; in 200 years, 1024 dols.; and this he makes the cost of production as thus calculated.

He goes on to say:—' Assuming that one hundred and eighty years are required for the growth of pine timber in the north-west part of the United States, these figures would seem to show that the pine forests of the United States are bringing to hand trees sold and consumed at a price very much below their actual value.' I know that the same thing has occurred in other countries besides America, and that in many cases the tree was worth to the country, as a tree, a great deal more than the price obtained for it as wood!

In the narrative of what I saw in sailing from Christiansand to Christiania, I have intimated that close upon the level of the sea, on the shores of the Skager-Rack, and on islands there, and the coast of Bohu's Bay, and of the Christiania Fiord, the coniferæ give a character to the scenery. But, as has been stated, these are not the only kind of trees found indigenous in Norway. There has been given on no mean authority a statement to the effect that, in Norway, journeying from north to south, and in some parts of the country in descending from a great elevation to the valley or plain, we pass through successive zones, which have been characterised as the zone of perpetual snow, where only a few ice plants, lichens, and mosses grow; the zone of the willow and birch; the zone of the pine and fir; the zone of the oak; the zone of the beech; and the zone of the cultivated fields. The statement may be useful as a memorandum; but, if great stress be upon it, it may break down. I have only found it true to the extent that there are here birch woods, or forests, reaching further north than any forest of pine or fir, though trees of these may be found in the same latitude; that willows are found in higher latitudes than these; and that there are forests of fir and pine further north than oaks; and oaks further north than beeches; and beeches growing further north, and at a high erelevation, than general cultivation has extended. But beyond an indication of these facts, I think the statement cited may mislead.

To the northernmost zone is assigned, as a character, its production of the birch, *Betula odorata*, Bechet. It flourishes throughout the whole country; but it is in west Finmark that it appears forming veritable forests; elsewhere it is found, as a tree which delights in the light, very frequently dispersed over clearings in pine and fir forests, and besides, on lands completely cleared of woods. The altitudinal limits of this tree are in Southern Norway about 1,100 metres; in the diocese of Drontheim from 600 to 700 metres; and in Finmark from 300 to 400 metres above the level of the sea. The birch is in Norway a tree of fine growth, and often takes an elegant form, with the exterior branches pendant in long clusters, measuring as much as 5 metres, or 17 feet. The birch may attain to a great age. It is not rare to meet with birches from 20 to 25 metres, 67 to 84 feet, in height, and 1·5 metres, 5 feet, diameter at the ground. The crown may stretch on all sides more than 10 metres, or 43 feet from the trunk. Besides the wood being used in carpentry, and as firewood, the external white bark is employed in the manufacture of a great many articles, and for roofing, being then covered with earth or turf.

This tree, and especially the form with long pendant branches, is the ornament of the valleys of Norway; certain specimens in isolated groups having acquired by their beauty a widespread fame, and are protected not only by the proprietors, but by the whole population of the valley, who are proud of them. They have generally particular surnames, borrowed ordinarily from the property to which they belong; one famous specimen is the *Holsbirk*, so called from the nature of the estate, Hols, Rennebu, in Orkedal, in the prefect of Sondre-Drontheim, 62° 58′ N.; the height of it is 25 metres, 83 feet 4 inches; the diameter is 1·09 metres, or 3 feet 4 inches, at 1·5 metres, 5 feet from the ground.

The willow of Norway, *Salix caprea*, known in Britain as the great round-leaved willow, is very much diffused over the forest regions of Norway, and even beyond the elevation above the level of the sea attained by the fir and the pine; and, as intimated, it may be seen to flourish in higher latitudes, or at least as far north as either of these trees. It grows at Hammerfest, 70° 37′ N., but only as a shrub or bush; and the allied tree, the aspen, *Populus tremula*, is likewise spread over the whole of Norway. In the south part of the country it may be met with, but in a dwarfed condition, at an altitude of 900 metres above the level of the sea. In the lower-lying country it attains a height of 30 metres, and even at Atlen, 70° N., in Finmark, it may attain to 18 metres, or 60 feet. Of Osiers, there are in Norway about 20 wild species, the greater part of them taking the form of bushes. They grow to a great altitude on the mountains, principally on moist slopes. In southern Norway, on spots a little sheltered, they may be found 1500 metres above the sea level.

GEOGRAPHICAL DISTRIBUTION OF TREES. 45

In the same zone may be observed the growth of the alder, *Alnus incana*, D. C., known in England as the hoary-leaved alder. It is met with very frequently so far north as West Finmark, where it may be seen with a height of 20 metres, 66 feet 8 inches, and a diameter of 30 centimetres at the level of the ground; and it grows to almost the same altitude as the beech.

The *Alnus glutinosa*, Gent., the common alder of Britain, is much less frequently met with. It flourishes always in low valleys along the banks of streams and a humid soil. It is scarcely to be met with above an altitude of from 250 to 300 metres; and its Polar limit is 64° N. lat.

Of the oak, two species are found growing wild in Norway, the sessile fruited oak, *Quercus robur*, W., and the common oak, *Q. pedunculata*, W. The latter is found on the west coast up to 63° N.; but the planted oak, up to 66° N. In the southern parts of the country it scarcely ever extends beyond 300 metres above the level of the sea. There are small forests of the common oak on the coast between Arendal and Flekkefiord, on the shore of the diocese of Bergen, and in Romsdal.

The beech, *Fagus sylvatica*, L., when planted, flourishes up to Stegen in Nordland, 67° 56' N.; and at Drontheim, 63° 26' N., its seeds come to maturity. As a wild tree, it is not met with beyond 61° N. Even in the southern parts of the country it does not extend beyond 250 metres above the level of the sea. It forms small forests near Tonsberg, near Larvik, near Arendal, and a little north of Bergen, at Sæim, an annex of Hardanger, 66° 35' N. At this point is found the most northern forest of wild beeches in the world.

The elm, *Ulmus montana*, S., the wych elm of Britain, is met with up to the 67° of north latitude. Its limit of elevation is generally from 500 to 600 metres above the level of the sea. At Solvorn, near one of the branches in the interior of the Sogne fiord, it forms quite exceptionally a little forest; elsewhere it is met with in small clumps or growing solitarily. It may attain to 32 metres in height. In Finmark, at Alten, 70° N., elms are met with which have attained to 8 metres, or 27 feet, in height, and 40 centimetres, or 16 inches, in diameter.

The ash, *Fraxinus excelsior*, L., is found growing wild up to the Molde, in Romsdal, 62° 44′ N.; but planted, it grows very much further north, and it ripens its seeds even within the Arctic Circle. In the south of Norway it flourishes up to 500 metres of altitude, and may attain a height of 30 metres, or 100 feet; while here and there may be seen trunks which at 40 inches above the ground have a diameter of from 1·5 to 1·8 metres—from 5 feet to 6 feet.

The lime, *Tilia parvifolia*, Ehrh., known in Britain by the designation the small-leaved lime-tree, extends in the east of Norway to 61° N.; but on the west coast to 62° N. It affects principally low-lying countries; but isolated trees may be found up to 500 metres above the level of the sea. Planted, it grows up to 64° N., and even up to Stegen, in Nordland.

The maple, *Acer platanoides*, L., known in Britain as the Norway maple, extends in Eastern Norway, in Sondre, Guldbrandsdal, to 61° 25′ N., and to an altitude of 260 metres above the level of the sea, though in the south it scarcely extends above 300 metres. Planted, it is found to the top of the Ramenfiord, in Nordland, 66° 18′ N.

The hazel, *Corylus avellana*, L., is met with pretty frequently in a wild state on the low-lying lands to the south of the Droutheim fiord. Further to the north it becomes more rare. It is met with, however, up to Stegen, in Nordland, 67° 56′ N., where its fruit still ripens. In the south of Norway it grows to an elevation of 500 metres above the level of the sea.

GEOGRAPHICAL DISTRIBUTION OF TREES. 47

The mountain ash, *Sorbus aucuparia*, L., *Pyrus aucuparia*, Gaerton, is very frequently met with everywhere in Norway, and in an ordinary summer its fruit comes to maturity. On mountains it extends to the limits of the birch, where it stops. With its branches of red berries, which offer a food greatly sought after by birds in great numbers, it gives to the mountain slopes in winter a peculiar and animated appearance.

In supplying such details we are drifting away from the consideration of what are generally know as forest trees to what are generally known as wild fruits; and why not? With these may be mentioned several others.

The crab apple, *Pyrus malus*, L., is met with here and there in a wild state up to the island of Yteros, on the Drontheim fiord, 63° 49′ N.; in the south it does not extend above 500 metres of altitude.

The gean, or wild cherry, *Prunus avium*, is met with in the interior of the Sogne fiord, at Urnæs, where there is a small wood of it, 12 metres, or 40 feet high, and 30 centimetres, or 20 inches in diameter.

The bird-cherry, *Prunus padus*, L., is generally diffused throughout the whole of Norway, even to the Tanaelv; in East Finmark, 70° 20′ N., its fruit ripens. In South Norway it attains the altitude attained by the pine, and sometimes it extends to a higher elevation.

The blackthorn, or sloe, *Prunus spinosa*, L., is found in South Norway, up to 60° N.

The barberry, *Berberis vulgaris*, L., shows itself in a sub-spontaneous state in many places up to 64° N.

The gooseberry, *Ribes grossularia*, L., is found here and there in a sub-spontaneous state in low-lying countries up to 63° N.

The red currant, *Ribes rubrum*, L., is found pretty generally spread over the whole of the country up to the eastern frontier of Finmark; and on mountains it grows beyond the limits of the pine.

The juniper, *Juniperus communus*, L., generally presents

itself as a bush in shrub form, and from 3 feet to 6 feet 6 inches in height; but in certain plains up to Selbo, in the perfecture of Sondre Drontheim, it assumes a fine conical form, and the proportions of a tree; and it attains to a height of 12 metres, or 40 feet. It is found everywhere up to the North Cape. It climbs up the mountains, passing the limits of the birch, and attains, like some of the osiers, an altitude of 1500 metres above the level of the sea.

The following are the various times of the flowering of trees and arborescent shrubs reared in the environs of Christiana, 59° 55' N., 10° 50' E. of Greenwich :—

Alnus incana, J. C.	... April	6-10	Syringa vulgaris, L.	... June	2- 6
Corylus avellana, L.	... ,,	6-10	—— chinensis Willd	... ,,	10-14
Salix Caprea, L.	... ,,	22-30	Juniperus communus, L.	,,	4- 8
Ulmus montana, L.	... May	4- 8	Cratægus, sanggin. Pall.	,,	4- 8
Betula glutinosa, Wallr.	,,	14-18	—— sacantha, L.	,,	14-18
Acer Platanoides, L.	... ,,	14-18	Rhamus cathartica, L.	,,	8-12
Larix Europœa, C.	... ,,	20-24	—— Frangula, L.	,,	8-12
Quercus pedunculata	... ,,	24-30	Cytisus alpinus, Mell.	,,	14-18
Pinus sylvestris, L.	... ,,	24-30	Rosa canina, L.	... ,,	18-22
Abies excelsior, D.C.	... ,,	24-31	—— rubiginosa, L.	,,	26-30
Fraxinus excelsior, L.	... June	1- 4	Robinca pseudo-acacia	,,	26-30
Æsculus Hippocastanum, L.	,,	1- 4	Sambucus nigra, L.	... July	1- 4
Sorbus aucuparia, L.	... ,,	2- 6	Ligustrum vulgare, L.	,,	4- 8
—— hybrida, L.	... ,,	4- 8	Tilia parvifolia, Ehrn	... ,,	8-12

The forests in Norway, according to a statement by M. F. L. Marny, in a volume treating of the forests of Europe, are extensive; but they are to a great extent suspended along the Scandinavian Alps, which separate this country from Sweden. The birch reaches there an altitude of 365 metres. In the diocese of Bergen the fir has still the gigantic proportions seen in the forests of Switzerland and Germany; but more to the north its size is diminished to stunted proportions, and at the Polar Circle it has totally disappeared; whilst in Swedish Lapland it advances yet to two degrees beyond this. In Norway the birch serves as a ladder to vegetation; it is the measure of its energy, and marks by the different states

GEOGRAPHICAL DISTRIBUTION OF TREES.

through which it passes, in proportion as it rises in altitude, the degree of weakness of vegetative life. To the weeping birch succeeds the *betula acer*, which replaces the white birch; after which comes the birch of the prairies, which passes in its turn through different gradations of size, and which at the Polar Circle is nothing more than a stunted shrub, of pyramidal form, and covered with moss.

In a paper on European forests, which appeared in the *Building News* for 1878, it is stated:—' The total area of forest-producing land in Norway is computed at about 37,000,000 acres, but in this survey considerable so-called forest land consists of comparatively unproductive rocks, swamps, and moors. The pine and fir, even more than in Sweden, constitute the riches of the Norwegian forests. The Scotch fir is found up to the most northern latitudes, and grows there up to a height of 3,400 feet above the sea level. The spruce fir ceases near the Arctic Circle. The forests are principally situated in the east of Norway, near Christiania, Hamar, Trondhjem, and Christiansand. Those of Bergen have long since been exhausted. In the western districts of the country forests can hardly be said to exist. As in Sweden, strict forest laws are now in force, but the mischief done in former times by indiscriminate felling will take a long time to repair.

CHAPTER V.

CONDITIONS ON WHICH DEPEND THE DISTRIBUTION OF VEGETATION.

If a map of the world were coloured with tints representing the number of varieties, or species, or genera, of different plants, growing on different localities, it would be found that the tints representing many of these would be deep in certain localities and apparently shaded away in others—as if there were certain marked centres, where the greatest number of the genera, or of the species of different plants occurs, and the number of these diminished as we receded from these centres, ending, perhaps, in solitary representatives of them in some distant country.

Similar, probably, would be the result of so colouring and shading a map in accordance with the numbers of the plants of any one species growing in different localities. And this might sometimes correspond with the tinting and shading on the other map; but more frequently, perhaps, it would be very different. In some localities where there is a great variety of species, there may be a comparatively small number of individual plants of some or all of these species; in others, a great number of the plants, but all of one or of few species; and in others a great number both of plants and of species.

Professor Edward Forbes, a distinguished naturalist of great promise, who perished in all the leaves of his spring, considered that there were such centres as I have spoken of, and that in all probability these were the localities in which these several species of plants first appeared; that thence they spread; that in some cases thence they might probably have been traced continuously from the centres to the most remote regions in which they are found, spreading, as does the fairy ring extend itself, but with this difference that instead of dying out in the centre, and on the ground on which they first grew, they continued to grow, and to grow it may be with increased luxuriance, in the locality in which they have longest been produced and reproduced abundantly; and that in some of the outposts where they now grow isolated from their congeners, they may have been severed from the centres whence they sprung by geological changes, such as the elevation or the depression of intervening land.

I am aware that these views have been opposed by Schouw and others; but the facts, upon the observation of which they are founded, remain the same and unimpugned; it is these with which I am concerned; my allusion to the views of Forbes is only incidental; and they are cited in aid of my illustration. Schouw, it is stated by Professor Balfour, considered that the existence of the same species in far distant countries is not to be accounted for on the supposition of a single centre for each species. The usual means of transport, and even the changes which have taken place by volcanic and other causes, are inadequate, he thought, to explain why many species are common to the Alps and the Pyrenees, on the one hand, and to the Scandinavian and Scotch mountains on the other, without being found on the intermediate plains and hills; why the flora of Iceland is nearly identical with that of the Scandinavian mountains; and why Europe and North America, especially the northern parts, have various plants in common, to which they have not been communicated by human aids. Still greater objections to this mode of explanation, he thinks, are founded on the fact that there are plants at the Straits of Madgalhaeris, and in the Falkland and other Antarctic islands, which belong to the flora of the Arctic pole; and that several European plants appear in New Holland, Van Dieman's Land, and New Zealand, which are not found in intermediate countries. Schouw, therefore, supposes that these were originally not one but many primary individuals of a species.

Without homologating these views in any way I avail myself of this as a contribution in aid of my illustration. I find it equivalent to an allegation that there have been not one but many centres of distribution whence *species* of plants have been dispersed; and I find it at once equally convenient, and more in accordance with the views I entertained in regard to the successive development of vegetation upon the earth, to look at this fact through Forbes' theory applied to the genus, and through that of Schouw applied to species. And I may further remark that neither do the views advanced by Forbes and Schouw, nor observed facts, necessitate the supposition that the dispersion from a recognised centre took place in a regular ever-extending continuous circle. Even the fairy ring to which I have alluded, though maintaining generally the circular form, shows not a continuous regular circular outline; often it is broken and it assumes something of a horse-shoe shape, sometimes it bulges out in the form of an oval or an ellipse, and in almost every case the regularity of the figures assumed, whatever it may be, is broken by void species, or by projections, or by both; and in the wider field of nature we find on the ascent of a lofty mountain that there are linear zones of different forms of vegetation; while, from the summit, may be seen on the plain, well-defined patches of vegetation conformed to particular geological formations, curved and winding lines of vegetation lining the course of some river or streamlet, or it may be, in a well-defined straight line traversing the plain, and ascending and crossing the mountains beyond, covering, but confined to a geological dyke, of different material, filling a crack or rent on the superficial stratum of the ground—a phenomenon of not unfrequent occurrence.

In view of this the geographical distribution of all plants, including trees—solitary trees, and trees of dense forests, continuous or far apart, may occasionally be attributed to dispersed seeds having alighted on soil favourably conditioned and favourably situated for the vegetation of the trees there found.

Provision exists for the dispersion of seeds by the fall of the fruit, by the carrying power of the wind of rivers, and of the sea, by the adhesion of them to the hair of beasts and the feathers of birds, by the voiding of them uninjured by animals which have eaten the fruit in which they were produced, and by man, purposely, or without design, in his migrations and his journeyings. But of the seeds thus dispersed, millions, and it may be the vast majority, alight where there are awanting one or more of the conditions under which alone they can germinate, grow, flower, fruit, and bring that fruit to perfection; and if this be the case with one or other of these conditions, there the plant, herb, or tree will not be produced, and permanently retain possession of the place: only where everything is favourable can a foothold for the plant be obtained and maintained. I say obtained and maintained for even where a plant may succeed in obtaining a foothold, it may subsequently be displaced in the struggle for possession with other plants better adapted for growth in the existing condition of the locality at any time. The resulting geographical distribution of plants, herbs, or trees, in any land at any time is the result of the influence of two conflicting operations,— dispersion and repression acting like the centrifugal and centripetal phenomona of gravitation, keeping the planet and satellite in their orbits, and the stone in the sling.

But there is another aspect of the subject which is not without its interest to the student of botany, and in so far to the student of forest science, which is suggestive of differences in conditions in different localities or at different times. The vegetation of any region may be sparse, or it may be abundant; but also under either of these conditions the plants of any kind (*genus*) may be uniform or varied, nearly uniform, or greatly varied; presenting in the locality few or many different forms (*species*). There may be in any locality, as respects the former, abundant or sparse vegetation, but as respects the latter, a poor or a rich flora; profusion of individual plants, but few varieties of species; or comparatively few plants, but amongst these great variety.

I may state in illustration that at the Cape of Good Hope we have a rich flora, but a rich vegetation would scarcely be a correct description of what is here to be seen. Of coniferous plants, such as are the cabbage and the wallflower, there are 17 genera, and one of them, the heliophila, is represented by upwards of 60 species. There are upwards of 100 species of crassula. Of the fig marigold, or mesembryanthum, there are upwards of 300 species. Among Cape bulbs there are of iris-like plants alone upwards of 20 genera—all of them, I may say, represented by several species. Of Cape heaths there are between 300 and 400 species described. Of pea like plants there are upwards of 80 genera; and of some of these, from 25 to upwards of 100 species are described by Dr Harvey as flora capensis; and of compostae, or daisy-like plants, there are 154 genera, and of these there are 1000 different species.

We have here a rich flora, but it does not follow that there is a rich and luxuriant vegetation. Continuous turf is almost unknown; bush there is in plenty; and there are fields of cereals and mesembryanthema, but there is nothing like the luxuriance of vegetation to be seen in many a wood and thicket, and winding lane in Britain.

And different conclusions must be arrived at according as we may adopt one or other of different principles in estimating the richness of a district in its flora and vegetation—according as we look to the number of *orders* of plants growing there, or to the number of *genera* of plants belonging to any one or more orders, or to the number of *species* of any of these genera, or to the number of *specimens* of these genera and species.

By Schleiden, writing on this subject, it is remarked:— 'The information we have obtained in regard to the so-called *habitations*—the place of growth and the native country of a plant—has enabled us to give an orderly arrangement to our conceptions relative to the distributions of plants, but much remains intricate and unexplained.

'In considering this subject, there are two essentially different points which have to be distinguished.

'The heath plants occur on dry, sunny, sandy plains; they *extend* from the Cape of Good Hope, through Africa, Europe, and Northern Asia, to the extreme limits of vegetation in Scandinavian Siberia; these plants are *distributed* on this great region in such a manner that South Africa has innumerable distinct species, of which, however, never more than a few individuals grow side by side; and that then, towards the north, the number of species suddenly diminishes in an important degree, while the number of individuals increases—the common heather, *Calluna vulgaris*, overspreads whole countries in millions of single individuals.

'In the first place, we readily see that only the first determination, that of the *occurrence*, relates necessarily to each individual; while, on the contrary, the *range of extensions* and the *mode of distribution* indicate courses which have scarcely any importance in reference to the single individual, but very great in relation to the larger groups of plants which we call species, genus, &c.

'From this it follows that the former only—the occurrence of plants—is related wholly, while the other two are related but partly, to conditions explicable by physical influence; yet we must, at first keep more to that arrangement, since it is strictly logical, which will remain fixed for incalculably long time, while of course the last only holds good for the existing condition of science; and when we review the various influences upon which the life and healthy vegetation of a plant are, according to our present physiological knowledge, dependent, we quickly find that only a small number of physical forces are as yet detected by us in their action on the organism; and that, on the other hand, a proportionately large number baffle our endeavours after a more accurate comprehension of their action, although we may safely assert that the life of the plant is, and must be, as much dependent on them as upon the others.'

He goes on to say:—'The operation is as yet a complete mystery to us, and many of the at present wholly incomprehensible conditions in extension and distribution of species may sooner or later find sufficient explanation in these influences.' Since this was written, the carefully made and well digested observations of Darwin have thrown no little light upon the subject, and introduced a method of study which is likely to throw yet more upon the subject in all its bearings.

'We find again,' says he, 'indications of the undoubted fact that the distribution of all plants is naturally regulated by law, but what laws we cannot evolve, and it looks as if it were wholly the result of caprice that particular plants are distributed widely over the globe, while others must be cribbed in the narrowest spot, as, *e.g.*, the *Wulfenia* found only on the Corinthian Alps; that particular families like the *Compositae* flourish abroad over the whole earth, while others like poppies and the palms only occur between very definite degrees of latitude on either side of the equator, the *Protoceae* only in the southern hemisphere, the cactus tribe only on the western half of the earth.

'Just as inexplicable is the *mode of distribution* of the families of plants. While the palms diminish in numbers from the equator into a higher latitude, the *Compositae* attain their highest development in the zones of mean temperature, their number of species diminishes from these in both directions equally towards the equator and towards the pole; while finally, the grasses increase constantly from the equator towards the pole.

'Having spoken of the increase of grasses as we proceed from the equator towards the poles, it is necessary that I should explain the mode of consideration according to which the distribution of the families is usually determined. Of sedges, to take that family as an illustration, the species found in the flora of France amount to 134; the species in the flora of Lapland, on the contrary, only to 55. France is therefore unquestionably richer in species than Lapland.

'But the matter stands in a somewhat different light when we consider these plants in relation to the total vegetation of the two countries; and since it is by this means we come to comprehend the characteristic of the region of vegetation, we can only allow this mode of consideration to be valid. Now France possesses altogether about 4,500 phanerogamous plants, and the sedges constitute only 1-27th of these; the phanerogamia of Lapland are confined to some 500 species, and 1-9th of these are sedges. The sedges are therefore a much more essential part of the Lapponic flora than of the French, the former having a relatively larger number of species than the latter. And this it is alone that is understood by the increase of species in a given direction.'

CHAPTER VI.

CONDITIONS AFFECTING THE GEOGRAPHICAL DISTRIBUTION OF PLANTS AND TREES IN NORWAY.

In Norway, as is the case everywhere, the distribution of forests, and of different kinds of forest trees, is effected, and I may say determined, by conditions varying in different districts, predominant amongst which are heat and moisture, and wind and soil.

From a knowledge of the natural history of different kinds of trees there may be learned much in regard to these conditions in the several localities in which they grow, much more than can be stated and proved in a few words, for it is not simply the sum of different measurements such as are applicable to these conditions, but the reflex and complicated influence of one and another of these upon one and another of them, upon which the ultimate result depends,—and on variations in some of them at different periods, embracing, amongst others, the periods of the germination of the seed, of the expansion of the leaf, of the formation of the flower, of the casting of the pollen, and of the maturing, and the falling of the fruit. By stating, however, the characteristics of different localities in connection with these conditions, there may be made a contribution towards a collection of facts which may be interesting to the general reader, and not without value to some youthful student of forest science; and this I shall attempt, citing and translating facts embodied in the report by Dr. Broch, to which I am indebted for much of what I have advanced in regard to the geographical distribution of the trees.

'Speaking generally,' says Dr. Broch, 'the Norwegian flora presents little variety. The gneiss, granite, and other like hard rocks, which constitute the principal mass of the ground present, like the clay, a vegetation somewhat uniform, essentially composed of a small number of species combined and diffused in immense masses.

'But it is not the same in the plain, where the earth is more friable. In the high mountainous lands there are often met with great and small oases of friable schists which often present the appearance of flower gardens in the midst of rocky deserts covered with mosses and heaths. It is there that the *Arctic flora* has its principal domain. Resembling greatly that of Greenland and Spitzbergen, it is specially characterised by species of plants diffused in great quantities, such as the *Dryas octopetal*, L.; the *Salix reticulata*, L.; the *Carex rupestris*, All.; the *Thalictrum alpinum*, L., &c. This schistose flora has everywhere a character pretty easily recognised. It does not stand mild winters. There, where the mountains are too near the sea, the schist is poor in species, and these Arctic oases of flowers are met with, and found to abound, in those regions in which the highest mountains, and the most extensive névés of snow, protect them against the hurtful sea air brought by the west winds.

'It is thus that a colony of Arctic plants shows itself to the east of Folgefoun, in Hardanger. In Lem, in Bage, and on the Dovrefjeld, and in its mountains, there is found a rich Arctic flora protected against the influence of the sea by the névé or snowfield, and by the Joetunfjelde. In the north it is only in Salten and in Swedish Lapland that a characteristic Arctic flora is to the same degree protected against the sea winds by the high mountainous regions, and the great névés of Sulitelma. In Finmark, in fine, is found a rich Arctic flora to the east of the high mountains and the great névés of the peninsula of Lyngen.

'Very different from this is the *Boreal Flora*, which equally shrinks from the climate of the coast, although it is found to the west and to the east of the mountain ridges. It is on the great masses of stones at the base of precipitous mountain walls that it is met with in the lower lying regions of the south. In the midst of hazels, elms, limes, and other broad-leaved trees, one often meets with a rich boreal flora, and this equally with the former has a uniform character. A group which approximates the boreal flora, but which is very distinct in its development, is the *sub-boreal flora*. In all the lowest lying parts of eastern Norway there is found from the level of the sea to an altitude of from 300 to 500 metres, and principally on the calcareous and dry schistose mountains of the Silurian formation, numerous species of plants which are neither met with in the higher regions nor on the west sides of the mountains.

'Further, the Norwegian flora has three other distinct groups of plants which depend on the humidity and on the climate of the coast. One of these, the *Sub-Artic*, is developed in the shady valleys there, and on the mountain expanses, kept moist by the melting of the névés. There prevail the *Archangelica officinalis*, the *Mulgedium alpinum*, the *Aconitum septentrionale*, the *Ranunculus aconitifolius*, the *Valeriana sambucifolia*, the *Struthiopteris germanica*, and many other plants. These have no dread of the coast.

'The second, the *Atlantic group*, seeks by preference the humid districts of the littoral of Bergen. It comprises— the *Ilex*, the *Digitalis*, the *Erica cinerea*, the *Bunium flexuosum*, the *Hymenophyllum Wilsoni*, the *Hyperichum palchrum*, the *Polystichum oriopteris*, and many others.

'The third, the *Sub-Atlantic group*, affects by perference the lowest lying districts and the most southern of the diocese of Christiansand. It comprises the *Gentiana pneumonanthe*, the *Sanguisorba officinalis*, the *Petasites alba*, the *Teucirum scorodonia*.

'The Norwegian flora, it thus appears, is composed of many groups of plants, each of which group is composed of species which have a somewhat, and pretty striking analogous development. But the species constituting these groups are never met with intermixed; and one may find in the same district, for example, near Christiania, which has a flora very rich in these representatives of the greater number of these groups; but in certain places they present themselves in a number of species and of individual plants giving to the vegetation of the place its particular character.

'When on a map there are marked the districts in which the different groups of plants principally predominate we discover that with the one exception of the most common species the development of groups is not continuous. The Arctic flora only shows itself it its characteristic form, here and there, in scattered colonies, separated by extensive species in which the Sub-Arctic predominates. The boreal flora also presents itself scattered over the low-lying districts of the eastern portion of the country around the Christiania fiord, and the Miosen lake, then at a distance to the west, beyond the Langfjelde, near the inland branches of the fiords of the western littoral, and lastly on the low lying lands on the north of the country. On the intermediate low lying coasts the Atlantic flora reigns. This is equally with the others isolated by the greater part of the Atlantic species finding themselves in the south-west portion of Sweden, but not around the Christiania Fiord. Some which have only been found on the west coast of Norway, find themselves again only on the west and the south coast of the German Ocean.'

Dr Broch goes on to say:—'The present flora of Norway cannot have always existed there. At a geological period, which is not very distant, the peninsula, up to the most remote plateaux of the land, was covered with a bed of ice and snow, overlooked only by the bare summits of some few of the most elevated mountains. At this time the trees, shrubs, and plants, which now beautify the valleys could not have lived. They existed, however, anterior to the so-called glacial period on other parts of the terrestrial globe. They have been found in a fossil state in beds of coal deposited prior to this period. It is an immigration from other countries which has produced the existing flora by degrees as the melting of the ice went on. This fact is finally established by the circumstance that Scandinavia—at least, in what relates to plants of higher orders—has scarcely one certainly distinct species which is entirely awanting in other countries.' The statement I give is that of a careful and accurate statistician resident in the country, with every source of information open to him, and every facility for forming a correct judgment at his command.

Special attention has been given by Count Saporta, and by the late Professor Heer, of Zurich, to the indications of the primary migration of plants in and from the North of Europe preserved in fossil remains: the former following up and expounding the views of the latter with all due honour to him, respectfully taking the position of a sympathetic approver and expositor of his views. According to them, vegetation originated in the Arctic Circle, spread southwards as the earth cooled, extending to the tropics—the hardier plants retained possession of boral regions, when others disappeared under a reduced temperature, and such alone it may be returning and regaining possession of the ground subsequent to the wide spread destruction of vegetation in the north in the glacial period. These views, and facts upon which they are based, I have detailed in a volume entitled *Forest Lands and Forestry of Northern Russia* (pp. 192-235).

CHAPTER VII.

TEMPERATURE.

OF the several conditions upon which depends the geological distribution of plants that which is most easily recognised, and most generally observed, is temperature. There are hardy arborescents, such as the gooseberry, the currant, and the cherry, which never fruit in warm countries; and there are tropical and sub-tropical fruit trees such as the bread-fruit, the love apple, the banana, and the grape, which may fruit with us in a hothouse, but not in the open air. During the meeting of the British Association in Aberdeen, in 1859, at which time I held the Chair of Botany in King's College, I was visited by a Polish savant, Professor Bialoblotsky, who was desirous of particular information in regard to the temperature of the district. Having stated generally his enquiry, he said:—'I have observed indications of the prevailing temperature. I see the furze flourishes luxuriantly; and I know of course that the cold is by no means severe. I have also remarked that the walnut grows well, but does not fruit; and I know thus something of the limit of heat in summer both as to degree and duration. But there are several details in regard to the duration of intermediate temperatures, and of the maxima and minima of which I wish to be informed, and also details of conditions under which they occur.'

So may one learn elsewhere much in regard to temperature from observation of the vegetation.

The following statements by Schleiden, formerly Professor of Botany at Jena, and afterwards at Dorpat, taken from his volume entitled *The Plant: a Biography*, supplies interesting illustrations of what has thus been advanced:—

'If from the snow-covered ice plains of the extreme north, where the red snow algae alone reminds us of the existence of vegetable organisation, we turn towards the south, a girdle first expands before us, in which mosses and lichens clothe the soil, and a peculiar vegetation of low plants, with subterranean perennial stems, and generally large handsome flowers, the so-called Alpine plants, gives a special character to nature. Almost all the plants form little flattened separate tufts; *Pyrola, Andromeda, Pedicularis, Cochlearia,* poppies, crowfoots, and others, are the characteristic genera of this flora, in which no tree, no shrub, is found. Leaving this region, which botanists call the region of mosses and saxifrages, we go southward, and at first we see little low bushes of birches, then more compact woods, into which the pines and other coniferous trees assemble, and we at last find ourselves in a second great zone of vegetation, which is characterised by the woods consisting almost exclusively of conifers, which thus impress a peculiar character upon the flora; firs and pines, Siberian stone pines and larches, form great widely extended masses of forest; by brooks and on damp soil occur the willow and the alder. On hills grow the reindeer lichen and the Iceland moss. In the cranberry, cloudberry (*Rubus Chamaemorus*), and the currant, nature gives spontaneously, though sparingly, food; and a rich flora of variegated flowers serves for the decoration of the zone, which stretches in Scandinavia to the northern limit of the cultivation of wheat, but in Russia and Asia almost to Kazan and Yakutsk; this we may call the zone of the conifers. To the south of this zone in Norway, so far north even as in the neighbourhood of Dronthiem, lat. 64° 15', the culture of fruit begins, though sparingly; soon appear the sturdy oak and woods of beech. In about the latitude of Frankfort-on-the-Maine, 50° 9', another tree joins company, which, in its bold picturesque mode of branching, takes its stand beside the oak, which in the beauty of its foliage, as well as the utility of its fruit, it far surpasses, the noble chesnut.

'The Pyrenees, the Alps, and the Caucasus, form the southern limit of the zone, in the more eastern portion of which the lime and the elm contribute greatly to the composition of the forest. In the hop, the ivy, and the clematis, we find here the first representative of the tropical climbers. The smiling green of the meadows alternates with the gloomy shadows of the forest, and man has taken possession of the earth, restricting the wild vegetation of the earth to that absolutely needful for wood and hay, and rich crops reward his industry.

'Leaving this zone of the deciduous woods to scale the rocky barrier of the Alps, suddenly there appear quite different plants; with the great woods of trees, the coriaceous shining leaves of which last through the mild winter, and round the mighty slopes of which climb the vine and flame-coloured begonias, unite the similar bushes of myrtle, tinus, arbutus, and pestialico [pestalozzia?]

'Here and there the dwarf palm is met with; labiate plants and crucifers, and fair-flowered rock roses replace in summer the spring flora of scented hyacinth and narcissus; but rarely, even on the most favoured spots, is the eye dazzled by the brilliancy of evergreen leaves, or the glaring play of colour on the naked, jagged mountain chains, or gladdened by the mild radiance of verdant meadows. In recompense, mankind has, in this zone of evergreen woods, seized upon the first of the Hesperides. It is

> 'The land where the citrons blow;
> Through the dark-green leaves the gold oranges glow.'

'Crossing the narrow straits of Gibraltar, and following the western coast of Africa, we soon reach the Canary Islands, on which we find that around sycamores twine mighty cissus stems, while capers and bauhineas interlace in the thickest of balsamic shrubs. The slender date-palm soars aloft, and the baobab grows up into gigantic masses of wood. The wondrous cactus-like forms of the leafless spurges, distinguished by their poisonous or pleasant-flavoured sweet milk, as the case may be, manifest a peculiar formative power in nature; and the dragon-tree, in the garden of Crotava, in Teneriffe, a gigantic arborescent lily plant, shows itself the growth of thousands of years.

'We have thus passed through six different zones of vegetation, in which we have found the continually increasing temperature calling forth ever a different, and ever a more luxuriant vegetation.

'If, in the warm region we have reached, we ascend one of the mountains found there—say the Pic of Teyde—we shall find that we pass through similar zones, but in order they are reversed.

'Man has taken possession of the soil of the plain at the base of the mountain, and dislodged the original vegetation. Through vineyards and maize-fields we ascend, till the shades of the evergreen bay-laurel surround us. Trees of the lace-bark tribe and similar plants succeed; we wander for a time through *a zone of evergreen forest trees*. At a height of 4000 feet we lose the plants which had so far accompanied us.

'A very small number of peculiar plants mark a quickly traversed *zone of deciduous trees*, and we come among the resinous trunks of the Canary pine.

'A *zone of conifers* shields us from the sun's rays up to a height of 6000 feet; then the vegetation suddenly becomes low; from humble bushes it passes into a flora which has all the characteristics of the Alpine plants, till finally the naked rock sets a limit to all organic life, and no snow and ice bedeck the summit of the mountain, only because its height of 12,236 feet does not, in a position so near the tropics, extend up to the region of eternal snow.

'Counting by the limits of vegetation, we have resurveyed, in a few hours' climb, the wide way from Spitzbergen to the Canaries—an extent of more than fifty degrees of latitude.

'In the whole way, downwards towards the south, and

upwards towards the summit of Teyde, vegetation varies conformably with the climatal conditions; and we can almost account for the observed distribution of plants by the mere increase and diminution of heat. If we extend our research further, we can even name particular plants which are found in particular northern latitudes, and in lower latitudes again regularly occur at particular heights on the mountains.'

The geographical distribution of trees in Norway, which has been detailed, is to a very great extent the result of difference of temperature; and that this is the case will probably be admitted by all. But under this general conclusion there are numerous phenomena, causes, effects, conditions, and results included. The results are the results of temperature under varied conditions, including amongst others, distribution in time, atmospheric pressure, and cloudy and cloudless skies. I have no intention to discuss here phenomena of meteorology and their effects on vegetation, and the effects of vegetation upon these. My present purpose extends only to the repeating of information collected in regard to facts and phenomena, and like the bricklayer's labourer, handing these over to others to make of them what use they may. But for the further information of the general reader, I may cite some views in regard to the process of vegetation which I think may be interesting, and which I consider substantially sound.

It is a generally recognised fact that heat is necessary to vegetation, and that without this, differing in the case of different plants, they will not grow, or if they grow they do not flower, or if they flower they do not fruit, or if they fruit they do not mature their seed, or if they mature their seed, the seed does not germinate and reproduce the plant. But to give explicitness to such statements it is necessary to distinguish things which differ. The average heat of the summer is something different from the average heat of the year; the average heat of the summer, and the average greatest heat of the summer, are not the same thing; the average cold of winter, and the average greatest cold of winter, may be very different; and the greatest degree of heat, or the greatest degree of cold, is not indicated by either the average greatest heat or the average greatest cold. A very much higher temperature than the average may be necessary to the maturing of fruit or seed, and the occasional occurrence of of a degree of cold, considerably in excess of the average degree of cold, may prove fatal to plants which could have withstood even a lower average cold, without detriment.

It may be accepted as a general statement that heat promotes vegetation, and that cold itself, a negative thing, operates negatively in checking it, inasmuch as it is an absence of the heat requisite for vigorous vegetation; but greater cold, amounting to frost, may prove positively destructive by freezing the sap, which in its expansion bursts the sap vessels, as the expansion of water freezing in pipes bursts these pipes. A heat in excess of the range of temperature within which a tree flourishes may limit its reproduction by seed by causing it to go all to wood; and a degree of cold in excess of that range may limit its capabilities of bringing its seed to maturity, producing, it may be, only a stunted growth; while an excess of cold beyond what would be followed by such a result may, as has been stated, kill the tree by the rupture of sap vessels through the expansion of the sap in freezing.

Dr Broch has collected details of observations of the temperature of Norway in respect of all of these particulars. He states that it is only of late years that observations such as are desired have been made. There exist for certain places a complete, long-continued series of observations. But these stations were limited in number; and though sufficient for the localities at which they were made, and as such valuable, the geographical distribution of these stations was not the one most favourable for obtaining the general results desired; and some of them do not even embrace all the elements of the climatic conditions of the locality.

In the close of 1860 there were established five stations for meteorological observations on the coasts of Southern Norway, and some years later another was established at Dovre, in the interior of the country. In 1866 there was organised a Meteorological Institute; and additional stations were established; and now *fifty-five* Norwegian stations send regularly their observations to the Institute. They are pretty equably distributed over the country, principally along the coasts; and several of the most recently established of them have communicated observations of no small value in their bearing on the climatology of Norway.

Meteorologists have on maps connected with a line places at which the mean—or, as many would call it, the average—temperature is the same. The lines thus produced are called isothermal lines—lines of the same heat. 'It is found,'—I quote Dr Balfour—'that while at the equator these correspond nearly with the lines of latitude, as we recede from the equator the two are widely separated. They run in curves, rising in their course from the east of America towards the west of Europe, and sinking towards the south in the interior of Europe. The yearly isotherm of 50° passes through the latitude of 42° 30' on the east of America, 51° 30' in England, 47° 30' in Hungary, and 40° in Eastern Asia. This want of conformity between the isothermal and latitudinal lines will be easily understood when we consider that a place having a moderate summer and winter temperature may have the same annual mean as one having a very cold winter and a very warm winter.' But even this isothermal curve is not uniform: it is marked by numerous minor curves.

According to Dr Broch, the isothermal lines in Norway follow as a general rule the configuration of the coasts, being affected, on the one hand, by the temperature of the ocean, and on the other side by that of the lofty mountain masses and chains of mountains, which follow the general direction of the shore.

The central point of the lowest mean annual temperature in the peninsula of Scandinavia is at a point intermediate between the Varanger fiord and the Gulf of Bothnia. There, as in Finnish Lapland, the mean annual temperature is as low as $-3° = 26·6°$ Fahr.

The yearly isotherm of $-2° = 28·4°$ Fahr., include in Norway the south part of the parishes of Kautokeino and of Karasjok, in the interior of Finmark. Farther to the south, the miners' village of Roros and its vicinity has also that mean temperature.

The isotherm of $0° = 32°$ Fahr., passes to the north of the Bay of Varanger, embracing the whole of Varanger, and the whole interior of the prefecture of Finmark; thence it curves towards the south-west, following generally the chain of the Kjoelen to the prefecture of North Drontheim, comprising of the prefecture of Tromso only the most elevated mountain parts situated towards the Lake Altevard. From the boundary of Nordland and North Drontheim it bends to the east, enters Sweden, and finally directs itself by a slight turning towards the bottom of the Gulf of Bothnia, whence it goes on to the east in Russian Finland. Further to the south, the isotherm of $0° = 32°$ Fahr., circumscribes in Southern Norway an oval space, comprising the south-east portion of the prefecture of Sondre-Drontheim, the north part of the valley of Osterdal, so far as the northern extremity of the Lake Stromsjoe, in the valley of Rendal, and lastly, towards the west, the Alpine region of Rundane.

Nearly 18 per cent. of the whole area of Norway has a mean annual temperature not exceeding $0°$, or $32°$ Fahr., the freezing point of water.

While the isotherm of $0°$ C., or $32°$ Fahr., encloses thus two separate portions of the country, the isotherm of $+2°$ C., say $36°$ Fahr., forms a single continuous line. It takes its rise to the north of the North Cape, and proceeds along the shore of Finmark by the outlying islands. It then traverses the prefectures of Tromso and of Nordland on its way towards the interior of the country, about midway between the coast and the mountain chain of Kjoelen. In the prefecture of North Drontheim it bends further inland from the coast, so that the upper part only of the valley of Namdal and of the Findlider finds itself on the east side—the coldest side of this line. Touching then the eastern frontier of the kingdom to the east of the lower part of Drontheim fiord, the line inclines toward the south-west, and encircles the Drontheim and Dovre fjelds. Thence it follows the line of the watershed of the Langefjeld, keeping in the same direction by places a little to the west, on to Œver-Telemark, the higher mountain of which it slightly grazes. Then it passes off, with a general direction towards the north-east, and almost in a straight line to the north of Lakes Kroederen and Speriten; cuts the Randsfiord and the Miosen a little to the south of the church of Elverm, in Œsterdal, to pass on into Sweden towards the east. It continues its principal direction towards the north-east, until it crosses the Gulf of Bothnia at its contraction between the Umea, in Sweden, and Carlely, in Finland.

In Norway this isothermal line comprises about 140,000 square kilometres, having a population of about 190,000 inhabitants.

Around this isotherm of $+2°$ C. there group other isotherms of a higher temperature. These succeed one another at short distances on the west side, but at much greater distances on the east side. On the north coast the colder air from the Arctic Ocean forces the isotherms to deviate a little more from each other on the two sides than they do further to the south.

The isothermal line of $4°$ C., say $39°$ Fahr., follows the coast line from the prefecture of North Drontheim to the valley of Romsdal. It passes along then within the coast, at a distance of from 30 to 40 kilometres. It bends inwards in passing all the great fiords, and follows the general direction of the coast to the Christiania fiord, whence it goes off to the east a little beyond Christinia, and enters Sweden.

The isotherm of 6° C., say 43° Fahr., follows a like direction to the middle of the Christiania fiord, where it diverges to the south-east, and passes thus into Sweden.

The isotherm of 7° C., say 45° Fahr., follows the coast line from Cape Stat, inflecting a little in the Sogne fiord, and again much more to the head of the Hardanger fiord; it cuts the Jæderon and continues to the Naye, to follow thence the coast, passing hard by the town of Arendal, whence it takes a south-easterly direction, traversing the Kategat, Southern Sweden, and the Baltic, to the coast of Pomerania. Thus it may be seen that the shore of Norway, from the Drontheim fiord to the Christiania fiord, has a mean average temperature of from 6° to 7° C., or 40° to 45° Fahr. In some favoured spots on the Hardanger fiord, and in the counties of Jæderen and Lister, the mean annual temperature is higher, ranging from 7° to 8° C. The mean annual temperature of Norway may be estimated at 2·5° C., or 36·5° Fahr. The mean annual temperature of Christiania is 5° 16′ C., or 41·25° Fahr.; in the course of the last forty years it has varied from 4° to 6·5° C., or 39·2° to 43·7° Fahr.

RAINFALL AND MOISTURE.

Not less marked than the influence of heat on vegetation is the influence of moisture. We have found, as may be found to be the case in Norway, and it is the case in many other lands, that many of the forests exist, and many of the forests of particular kinds of trees exist, only along the banks of rivers. It may be the case that, besides the moisture existing in the soil and in the atmosphere of such localities, there are other conditions of these combining with this to secure this result; but no one can doubt that most prominent amongst the conditions is the humidity which there prevails. Elsewhere much of the moisture which is required for vegetation is supplied by the rainfall; and to the measure of this in different localities attention is given by the students of Forest Science.

The distribution of rain in Norway, it is reported by Dr Broch, is in all seasons very much the same. The most rainy region comprises the shore and the fiords from Bergen to Cape Stat; the measure of the annual rainfall there is about 2 metres or 80 inches. At Bœmmel fiord the coast begins to rise, whilst otherwise its general direction presents conditions favourable to the production of rain; from this fiord to Cape Stat it consequently falls in greater quantity than elsewhere. To the north of Sogne fiord the Justedalsbrae, or braes of the Justedal, act as a powerful refrigerator, the west side of which is the most rainy region in Norway, while the country to the east of this is the most dry. On the coast of the Nordland the rains are also considerable in some places; the névé, or snow field, of Svartis, between Ranen and Salten, plays also there an important part in producing like results. Further to the north the temperature is so low that the quantity of rain falling there can never equal that of Southern Norway; nevertheless, the condensation of vapour occurring there with great frequency, the number of rainy days is considerable. On the coasts from the Romsdal to the Skudesnaes, and at Christiansand, the annual rainfall measures about a metre or 40 inches, while at Tromso and at Christiania it is only about a half of that; and on the mountains of the Dovrefjeld it is only about a third of that quantity.

When studied in connection with forestry, the distribution of the rainfall, both in time and in space, demands attention. In a volume entitled *Forests and Moisture; or, Effects of Forests on Humidity of Climate*,* I have shown that while the geographical distribution of the rainfall has to some extent determined the distribution of forests, one effect of forests, when created, has been to affect the local distribution of rain in time and quantity, and in space. In a well-wooded land the rainfall may be found to be diffused in showers over a great part of the year; while in a land otherwise under similar conditions, but devoid of forests and other vegetation, the rain falls only at distant intervals—months or years apart—and falls in torrents. And again, in the former case, the rainfall may be generally diffused over the whole area of the district; in the latter it may fall in torrents here and torrents there, leaving extensive regions unvisited by rain for long. Numerous

---

* *Forests and Moisture; or, Effects of Forests on Humidity of Climate.*—In which are given details of phenomena of vegetation on which the meteorological effects of forests affecting the humidity of climate depend,—of the effects of forests on the humidity of the atmosphere, on the humidity of the ground, on marshes, on the moisture of a wide expanse of country, on the local rainfall, and on rivers,—and of the correspondence between the distribution of the rainfall and of forests,—the measure of correspondence between the distribution of the rainfall and that of forests,—the distribution of the rainfall dependent on geographical position, determined by the contour of a country,—the distribution of forests affected by the distribution of the rainfall,—and the local effects of forests on the distribution of the rainfall within the forest district.—Edinburgh: Oliver & Boyd. London: Simpkin, Marshall, & Co. 1877.

instances of such irregular distribution and its consequences I have given in a volume entitled *Hydrology of South Africa*.*

Along the whole coast of Norway it is the autumn which furnishes the greatest quantity of rain; at Christiania it falls in the month of August. In the central part of the country the rainfall is least in spring.

The number of rainy days is generally proportionate to the quantity of the rainfall. Still the rainy days are relatively numerous in northern Norway, and likewise at Christiania. On the Dovrefjeld the days of rain in the course of the year are 90; on the coast of the Skager Rack the number is from 90 to 100; at Bodoe, Tromso, and Vardo, about 120; at Skudesnaes, at Christiania, on the coast of the Romsdal, and of Heligoland, about 140; at Bergen and in the fiords to the north of that town, and also at Vesteraolen, about 160; at the Lofoden Islands, 180. Thus it appears that in the interior of the country one day of rain in four may be counted on; while at the Lofoden Islands one day of rain in two, taking the whole course of the year, may be reckoned on.

Connected with this, there is another point deserving attention. In point of fact, though it may not be generally remarked, the light is one of the important influences promoting vegetation, as well as heat and moisture. And the number of cloudy days is closely related to the number of days on which rain falls.

It is found the effect of the so-called rainfall on the

---

* *Hydrology of South Africa; or, details of the former Hydrographic condition of the Cape of Good Hope, and of causes of its present aridity, with suggestions of appropriate remedies for this aridity.*—In which the desiccation of South Africa, from pre-Adamic times to the present day, is traced by indications supplied by geological formations, by the physical geography or general contour of the country, and by arborescent productions in the interior, with results confirmatory of the opinion that the appropriate remedies are irrigation, arboriculture, and an improved forest economy; or the erection of dams to prevent the escape of a portion of the rainfall to the sea,—the abandonment or restriction of the burning of the herbage and bush in connection with pastoral and agricultural operations,—the conservation and extension of existing forests,—and the adoption of measures similar to the *reboisement* and *gazonnement* carried out in France, with a view to prevent the formation of torrents and the destruction of property occasioned by them.—London: Henry S. King & Co. 1875.

distribution and growth of trees is different when the deposit is in the form of snow from what it might have been had it been in the form of rain; therefore this also demands attention.

At Vardo rain falls during 54 days, and snow during 71 days of the year. The Dovrefjeld has 41 days of rain and 48 of snow. In all other places at which observations have been made rain falls much more frequently than does snow. At Christiania the days of rain are double those of snow; at Bergen the number is five times as great, and at the Lofoden Islands 1·5 times greater.

As might have been anticipated, the greatest quantity of snow falls in the winter. The summer months are generally exempt from snow, even in Finmark; at Vardo there is no fall of snow in July and August. This is also the case at Bodo. At the Lofoten Islands, on the contrary, snow may fall any day, even in the months of July and August, snow falling in each of these months upon an average three times in ten years. In Southern Norway the months of June, July, August, and September are free from snow.

At Vardo, and on the Dovrefjeld the number of days of rain and of snow are about equal in May and in October; elsewhere it is so in the beginning of April and in November; and on the west coast it is only so in the month of March. At Bergen the number of the days of rain is in excess of that of the days of snow.

The quantity of rain falling in a day is pretty variable. Least considerable in the central part of the country, it is there calculated on the mean of the year, about 4 mm., ·16 inches; it increases towards the coast At Vardo and Bodo it is about 6 mm., ·24 inches; at Christiansand, Skudesnaes, and Sandoesund, from 7 to 8 mm.; at Bergen and Mandal, about 11 mm., ·44 inches. It is pretty nearly proportionate with the annual rainfall, and with the number of rainy days. We may thence evolve the general law that in these places in which there is the greatest abundance of rain, this is attributable to the combination of the three elements—frequency, force, and duration.

RAINFALL AND MOISTURE.

But it is not by rainfall alone that vegetation is promoted. Account should be taken of the dew, and not of the dew only, but of moisture in the atmosphere, which is to the eye invisible. In a very ancient record anent cosmogony we read:—'The Lord God had not caused it to rain upon the earth, and there was not a man to till the ground; but there went up a mist from the earth, and watered the whole face of the earth.' I cite this in illustration of the fact that, for a long time indeed, it has been a recognised fact that a mist or dew may promote vegetation. Dry as is a London fog, a Scottish mist is a connecting link between this and the drizzling rain. I have already had occasion to report the observations made on the frequency of mists at different seasons in different parts of Norway.

The mist, the cloud, the dew, and the rain, are all produced by a fall of temperature, rendering the air incapable of sustaining in its composition a quantity of moisture which was previously there, but invisible; and it may sometimes be more important to determine the humidity of the atmosphere than the rainfall, this being only a deposit of surplus moisture in excess of what the air could sustain in solution.

While the pressure of the atmosphere is measured by its equivalent in the weight of the column of mercury in the barometer, the quantity of vapour in the air is measured by its tension, which is also so determined.

Taking the mean of the whole year, the greatest quantity of aqueous vapour in the atmosphere of Norway is found on the coast from the embouchure of the Christiania fiord to the island of Karmoe, where its tension is 6·5 mm. A tension of 6· mm. extends over the middle of Christiania fiord, over the Boemmel fiord to Bergen, at Stat, and on the coast of Romsdal. The tension of 5 mm. passes a little to the north of Christiania, over the Inner-Sogne to the east of the Drontheim fiord, and away to the north, following the coast of Nordland, up to the Lofoden Islands. The vapour tension in the interior of Southern Norway, in the interior of Nordland, and on the coasts of the prefectures of Tromso and of Finmark, is between 5 mm. and 4 mm. In the environs of Roros, in Southern Norway, and in the interior of Finmark, it scarcely exceeds 4 mm.

The quantity of aqueous vapour appears thus to depend primarily on proximity to the coast, where it is greatest; but it depends also on the latitude and difference of temperature thus produced.

In the winter season the hygrometric lines are substantially similar to those of the mean of the year. In January the tension is that of 4 mm. at Christiansand, Bergen, Sondmor, Christiansand, and Bodo; 3 mm. in Smalenene, Inner-Sogne, Stjordal, and Sengen; 2 mm. on the Dovrefjeld, in the interior of Finmark, and in South Varanger.

In summer the quantity of aqueous vapour is much more considerable than in winter, and the distribution of it over the country is somewhat different. In July the line of tension of 10 mm. passes through Mandal and Tonsberg, and over this extent it is in the month of July that the greatest quantity of aqueous vapour throughout the year is found. The line of tension of 9 mm. passes outside the west coast from Skudesnaes to Cape Stat, and there inclines to the north-east, taking a direction to the east of Alesund, and towards Christiansand. Thence it inclines towards the south, descending towards Inner-Sogne, Hardanger, it proceeds towards the east to Drobak, and finally passes into Sweden. The great inflection of this line indicates the excess of aqueous vapour on the west coast relatively to the interior regions of the fiords in Sogne and Hardanger. These regions, viewed in regard to the hygrometric condition, approximate that of the interior, where the line indicates a region relatively deficient in moisture. The transition from the embouchure to the head of the Christiania fiord is very much marked; there the tension diminishes from 10·5 mm. to 8·6 mm.; it is the greatest variation seen in Norway. The line of tension of 8 mm. in July passes over the islands of Vestrealen towards the east. That of 7·5 mm. passes over Tromso and South Varanger.

The annual variations of the aqueous vapour tension accords well with those of the temperature of the air, and especially with those of the sea. At all the stations the greatest quantity of vapour is found in August, and the least in winter.

The most considerable variation of aqueous tension manifests itself on the Christiania fiord, on the eastern part of the country, and on the Drontheim fiord, where it amounts to 7 mm. The least variation is seen along the west coast where, from the Jæderan to Vadso, it keeps itself within 5 and 5·5 mm.; on the coasts of the Remsdal it is still less. In this respect the contrast between the continental climate and that of the coast manifests itself in the same way as the annual variation seen in the atmospheric temperature. On the high mountains of the Dovrefjeld both the quantities and the variations are less than they would be at the level of the sea.

The diurnal variation on the west coast is conformed to that of the temperature. At Christiania, on the contrary, the maximum tension is in the evening, which is especially remarkable in summer, when the diurnal variation is greatest.

The mean *relative humidity* of the year is greatest at Vardo, 85 per cent., and also on the coast from Mandel to Skudesnæs, being more than 80 per cent. It is a little less at Tromso, Christiansand, Alesund, and Bergen, 78 to 79 per cent., and it falls to 72 per cent. at Christiania. As a general rule, there the relative humidity is greater on the coast than in the interior of the country. It seems to be the same in Sweden, the coast of which, on the Baltic, shows a humidity very great compared with that of Norway.

In winter it is the west coast of Norway, from Cape Stat to the North Cape, which shows relatively least humidity, while this goes on augmenting steadily towards the Gulf of Bothnia. In January the relative humidity is at Tromso and at Bodo only 77 per cent., at Stat 80 per cent.; but at Christiania it is 88 per cent., at Vardo 90 per cent., and on the east coast of Sweden about 96 per cent. This proportion, which is the opposite of that of the whole year, is mainly attributable to the high winter temperature of the coast, and the low winter temperature of the interior.

In summer the shore from Vardo to Cape Stat, and to the Jæderen, has the greatest relative humidity, more than 80 per cent., while the central part of the peninsula has a pretty dry atmosphere, the relative humidity of which is only from 50 to 60 per cent.

The variation of this relative humidity is a little irregular on the west coast. It is, on the contrary, more regular to the east of the Naze, and in the diocese of Drontheim. The month of May is remarkable in Southern Norway as being the dryest of all the months. Everywhere the greatest humidity occurs in the winter season. On the west coast it is considerable in August, but diminishes again in October. In the western part of the country, and in the diocese of Drontheim, it has its minimum in winter and its maximum in winter without any marked interruptions in the regularity of its increase and diminution. The annual variation of relative humidity is greatest in the eastern part of the country, where it exceeds 30 per cent. It amounts to 20 per cent. in the Drontheim fiord; it is a little under 15 per cent. along the Skagerack, at Bergen, Inner-Sogne, and Fosen; and below 10 per cent. at Skudesnaes, on the coast of the Romsdal, at the Lofoden Islands, and on the coast of Finmark.

The number of foggy days varies with different localities. While at Vardo there are 18 in the year, in the Lofoden Islands 13, in Christiansand 7, and the Dovrefjeld 10, in Aolesund, Skudesnaes, Mandal, and Sandosund there are 20, in Bergen 40, and in Christiania.

And the annual variation is characteristic. In the

eastern part of the country, from Cape Lindesnaes to Christiania, the greatest number of foggy days occurs in winter; there are almost none in summer. On the western coast, on the contrary, and on to Vardo in the north, the prevalent fogs occur almost exclusively during summer. At Bergen they are distributed pretty equably over the whole year; but it is in June that they are most frequent. The foggy season then is in winter in the eastern part of the country; it is in summer in the western.

CHAPTER IX.

RIVERS.

I HAVE had occasion oftener than once in giving preceding details to make mention of the circumstance that many of the forests in Norway, and more especially the forests of broad-leaved trees, are situated in river basins, lining the river bed. In South Africa I have often seen that from some little eminence, whence the traveller could survey an extensive plain, one could trace the venation of it by river beds by the well-marked line of trees upon the banks; and I find a similar appearance is presented by the forest maps of several divisions of Norway.

It is on moisture in the atmosphere derived from the sea to a great extent by evaporation on which vegetation depends. Even when sustained in solution, and invisible, this is absorbed by the soil, which has a great affinity for it. The power of the air to sustain water in solution varies with varying temperature; and a fall in temperature may occasion a deposit of the surplus beyond what can be sustained, which deposit may, according to circumstances, take the form of cloud, of mist, of dew, of rain, of snow, or of hail. The soil, also, can only retain a definite quantity of moisture, varying with its constituents, and the excess which may reach it at any time passes off in streamlets, brooks, and rivers. In this respect the rain and the river are alike—they are the drainage off of water in excess; but in this also they are alike—they may convey moisture from where it is not required to some other place where it may be utilised in the promotion of vegetation. And all along a river from its rise to its flow into the sea, the reservoir of the world, the banks to a distance varying with circumstances are moist, though at a greater distance the ground may be arid and sterile. To this is attributable the growth of those trees in the situations indicated.

But these rivers have attractions for others besides students of forest science.

It is the wild and varied scenery produced by rock and river, fiord and forest, and mountain and lake, which makes Norway interesting to many British tourists, while its salmon fishings in rivers and lakes makes it also attractive to others.

'Of all the contrasts which Norway presents to other mountainous countries,' says Forbes in his volume entitled *Norway and its Glaciers*, 'the abundance of running water is perhaps the most striking to a stranger; its noble rivers, and its impressive waterfalls, are perhaps the features of the scenery most generally dwelt upon, and many tourists seem to make the latter the sole and main object of their search. This I think is a mistake. A cascade is a noble object as forming part of a landscape, but it is often situated so as to be well seen only when every other part of the landscape is excluded. If dwelt upon exclusively it becomes a mere *lusus naturae*, not an element of scenery; and if made the exclusive object of a laborious journey, it can scarcely fail to disappoint. I have not seen the most celebrated falls of Norway, for my other objects of enquiry did not lead me near them; but having visited those of other countries I have come to the conclusion that, setting aside the curiosity and variety of a lofty cascade, small waterfalls, unexpectedly discovered in picturesque situations, convey a truer sense of pleasure and beauty to the mind than the thundering shoots which tumble often into nearly inaccessible gorges. In the former class Norway abounds beyond calculation; running water of a bright and sparkling green is seen on every side, at least in the valleys; it pours over cliffs often in a single leap, but more frequently, and more effectively, in a series of broken falls, spreading laterally as it descends, and rivetting the attention for a long while together in endeavouring to trace its subtile ramifications. The sound is rather a murmur than a roar, so divided are the streams, and so numerous the shelves of rock tipped with foam; whilst a luxuriant vegetation of birch and alder overarches the whole, instead of being repelled by the wild tempest of air which accompanies the greater cataract. At other times single threads of snow-white water stretch down a steep of 2000 feet or more, connecting the fjeld above and valley below; they look so slender that we wonder at their absolute uniformity and perfect whiteness throughout so great a space—never dissipated in air, never disappearing under *débris ;* but on approaching these seeming threads we are astonished at their volume, which is usually such as completely to stop the communication from bank to bank.

'The source of this astonishing profusion of water is to be found in the peculiar disposition of the surface of the country. The mountains are wide and flat, the valleys are deep and far apart. The surfaces of the former receive and collect the rain, which is then drained into the narrow channels of the latter; and as the valleys ramify little, but usually pursue single lines, and are wholly disconnected from the fjelds by precipitous slopes, it follows that the single rivers which water those valleys represent the drainage of vast areas, and are supplied principally by streamlets which, having run long courses over the fjelds, are at last precipitated into ravines in the form of cascades. The system might be represented in a homely way by great blocks of houses in an old-fashioned town, the roofs of which collect and transmit the rain-water by means of communicating gutters, until, on reaching the street, the whole falls by means of open water-spouts, flooding the waterways below.

'But there is also another reason for this striking abundance of water. The fall of rain is large, if not excessive, over a great part of Norway. It is also, no doubt, greater on the fjelds than in the valleys of the interior. The height of the mountain plateau is such as to be covered more or less with snow during two-thirds of the year or more; during the period the rivers and cascades are, comparatively in many cases, absolutely dry. The vast accumulations of autumn, winter, and spring, are to be thawed during the almost constant warmth of the long summer days. In this season alone the interior of Norway is usually visited, and we see the result in the amount of drainage concentrated into that brief season. In the Alps, no doubt, a similar cause is active, but the comparative want of cascades is explained by the absolute want of table lands, and the infinitely ramified character of the valleys. In the Pyrenees, which have a still more ridge-like character than the Alps, the cascades are more numerous, but yet far more scanty.'

By Dr Broch it is reported in regard to the rivers of Norway—I give a free, but substantially correct translation of his statements:—

'From the position of the mountains the large rivers can only flow from north to south, or south to north. Affluents and small rivers on the west coast alone flow in a different direction. All the rivers of south Norway flow through a chain of lakes generally of considerable size, and almost always elongated in the direction of the river's course. They are often so narrow that they may be considered only expansions of the rivers where their depth, which is often very considerable, does not impart to them the character of lakes. These lakes form reservoirs which receive and modulate the flow of the rivers, and with a view to increase this effect there has been established moveable dykes at the embrochure of some lakes.

'The rivers of Norway experience a pretty regular flood, produced in the end of May and in June by the melting of snow in the valleys and on the elevated plateaux. The magnitude of this flood is very variable and depends not only on the quantities of snow which may have fallen in different localities in the course of the winter, but also, and perhaps still more, on the rapidity with which the snow is melted under the heat of spring and summer. In the southern portion of Norway the period of the summer stolstice in 1860 everywhere produced the greatest floods witnessed in the course of the century. On the high mountains the summer heat does not tell with effect till later, producing ordinarily towards the end of July a second, but lesser, flood. The great névés and the everlasting glaciers of the high mountains, the lower extremity of which is being continuously melted by the heat of the sun, produce flowing affluents in winter and in summer alike; the affluents from lower lying lands, on the contrary, diminish, and the lesser of them cease even to flow during the winter.

'In autumn again, towards the end of August, and in September, and during the first days of October, there is generally a flood, but of lesser magnitude, produced by the abundance of rain, or of snow which melts at once through the action of warmer currents of air. This autumn flood is ordinarily much less than the spring floods; but for all that it may assume great proportions. Thus, for example, in the beginning of the month of October 1795, the autumn flood in Southern Norway attained everywhere almost the same magnitude as the great spring flood of 1860.

'This spring flood is of great importance for the floatage of wood, which in the forest districts is carried on so long as the affluents have water enough for the purpose. An insufficent flood occasions great loss and damage to the proprietors of forests and to all engaged in timber trade. But a flood too strong may equally occasion difficulties in the floatage of the wood.

'In the western part of Norway the rivers are shorter and receive their affluents from the névés and glaciers. It is only in the valley of Romsdal, and in the two prefectures of Drontheim, that we meet again with large rivers. These have fewer lakes, and the flood is more violent, and often causes considerable destruction. The same may be said of the greater part of the rivers of the prefectures of Nordland and of Tromso.

' The Glommen is the greatest river in the Scandinavian peninsula, if we leave out of account the continuous or successive rivers in the water course of the Klarelv, the Venern, and the Gotelv, in Sweden. The Glommen has a length of 567 kilometres, and a basin of 40,400 square kilometres. Its most important affluent is the Vormen, which flows from the Lake Miosen—which has for its principal tributary the Guldbrandslagen, which has its source from the Lesjeskogsvand, on the line separating the Guldbransdal and the Romsdal. Where it joins the Glommen at Naes, in Romerike, the Vormen has almost equal dimensions with those of the Glommen—its length there being 322 kilometres, and its basin measuring 17,050 square kilometres, while the length of the Glommen above the confluence is 435 kilometres, with a basin of 19,880 square kilometres. . . . A little above its embouchure the Glommen divides into two branches, the larger of which falls into the sea, traversing the town of Fredrikstadt, the lesser, immediately to the west of this.'

Of the wood floated on this river I have had occasion to give some account in a preceding chapter [ante p. 10].

The Dramselv, the second river in Norway, passes through Drammen, and falls into the Drammen fiord, a branch of the Christiania fiord. It is formed by the confluence of the water course of the Rands fiord and the Bægna immediately above the Tyri fiord, in which the Dramselv has its source. Above that the water course of the Rands fiord. calculated by the principal river Dokka, flowing from the plateau of the Oplande, has a length of 141 kilometres, and a basin of 3710 square kilometres; while that of Bægna, which has its source in Filefjeld, has a length of 193 kilometres, and a basin of 4800 square kilometres. A little after its issue from the Tyri fiord the Dramselv receives its affluent, Hallingelv, from the Hemsedals fjeld and the Hallingsdal fjeld, with a length of 186 kilometres, and a basin of 5230 square kilometres. In itself the Dramselv has a length of 263 kilometres, and a basin of 16,890 square kilometres.

The rivers of Finmark, the Altenelv and Tanaelv—the latter of which is, after the Glommen and the Dramselv, the third greatest river in Norway—do not themselves expand into lakes, but they receive affluents from a great many small lakes, and they pass through some others. The river Patsjoka is the flow to Varanger fiord of the great lake Enare, measuring 2530 square kilometres, situated in Finnish Lapland at an altitude of 250 metres.

In my account of my visit to Christiansand mention is made of the Toppdal river, and the Torristal, or Otterdale river.

Of the Topdalselv, debouching to the east of Christiansand, it is stated that it is the discharge of the eastern part of the mountains of Sætersdal, and has a length of 136 kilometres, with a basin of 1,900 square kilometres.

Of the Otterelv, the river I visited, and which debouches into the western part of Christiansand, he states that it flows from the midst of the mountains of the Sætersdal, receiving affluents from the Byklefjeld, the most southern part of the Langfjeld, and that it has a length of 226 kilometres, with a basin of 3,660 square kilometres.

Of the waterfalls in Norway mention has been made again and again in preceding chapters. Of such the most remarkable, on account of the quantity of water, and at the same time, the most important, in consequence of its proximity to the sea, is the Sarpfoss, formed by the Glommen near its embouchure. The flow of this mass of water varies from 100 to 150 cubic metres per second in winter, and from 2000 to 3000 cubic metres per second when the river is in flood; and some years it has amounted to 4000 cubic metres. The mean flow may be estimated at 800 cubic metres per second. The perpendicular height of the fall of Sarpfoss is 21 metres.

CHAPTER X.

LAKES.

A MARKED feature of Norwegian scenery not to be overlooked is the lakes. There, as elsewhere, lakes are, as they are generally—I had almost said universally—a mere expansion of a river. Seldom, if ever, is its source only a spring in its depths. One or more rivers or rivulets flow into it and feed it, filling the hollow or valley to overflowing, and a different name may be given to the outflowing stream than that borne by any one of the feeders, but this does not affect the fact that the lake is only an expansion of one of these, and a receiver of its affluents at that place.

A lake known to most tourists in Norway is the Miosen lake, not far from Christiania, lying by the highway leading thence to the Dovrefjeld and the western coast. The Miosen lake is reached from Christiania by a railway which passes through a rich fertile valley, with a pretty river winding along it, and then plunges through some dense forests of tall pines, with stems so straight and uniformly tapering that they appear like huge fishing-rods. Their bark has a fine rich colour, which reflects the sunlight, and fills the whole atmosphere between the labyrinth of bare poles with a warm tinge, similar to that produced by stained-glass windows in the aisles of a Gothic cathedral.

The Eidsvold station of the railway is beautifully situated on the river which flows from the Miosen lake to the Glommen, and thence a steamer, formerly owned chiefly by one of our countrymen, famous in his day, by whom the railway was constructed, sails to Lillehammer, at the northern extremity of the lake. The Miosen is a long, narrow lake, not unlike our Windermere, but on a larger scale; being some seventy miles in length. The mountains that form its basin rise to a height of about 2000 feet at their visible summits; their form is not remarkable, but their sides, sloping down to the lake, are covered with rich emerald verdure, rivalling, if not excelling, our own green fields, and even those of Ireland. These slopes are backed by fine woods of birch and mountain ash, and dotted about them are the wooden farm-houses. Altogether, the Miosen is a beautiful lake, though not exciting rapture. About half way on the lake is the site of the ancient town of Stor-Hammer—*Störr* signifying large, and *Hammer* the same as our ham or hamlet. The ruins of its old cathedral remain, and near it, or, I believe, including it, is the farm of George P. Bidder, once the famous calculating boy,* and

---

* From time to time there have appeared such prodigies of boys who seemed not to calculate but to see, as many men see, that 1 and 1 are 2, that 2 and 3 are 5, what are the sums and products of numbers greater far than these. Such was Jididiah Buxton, such was Terah Colburn. The latter has left us a memoir of his life and achievements. On one occasion he was asked to name the square of 999,999, which he stated to be 999,998,000,001. He multiplied this by 49 and the product by the same number, and the total result he then multiplied by 25, the two latter operations being comparatively simple from the proportions which 25 and 50 bear to 100. He raised with ease the figure 8 to the sixteenth power. He named the squares of 244,999,755 and 1,224,998,775. He instantly named the factors 941 and 263, which would produce 247,483. He could discover prime numbers almost as soon as named. In five seconds he calculated the cube root of 413,993,348,677. But he admits that George Bidder was even more remarkable in some ways than he was; he could not extract roots or find factors with so much ease and rapidity as he, but he was more at home in obstruse calculations.

At three years of age George Bidder answered wonderful questions about the nails in a horse's four shoes. At eight, though he knew nothing of the theory of ciphering, he could answer almost instantaneously how many farthings there were in £868,464,121.

An octogenarian who saw these statements in the *Spectator* subsequently sent to that journal the following account of his recollections of two interviews which he had with him when Bidder was a little boy :—' In the autumn of the year 1814, I was reading with a private tutor, the Curate of Wellington, Somersetshire, when a Mr Bidder called upon him to exhibit the calculating power of his little boy, then about eight years old, who could neither read nor write. On this occasion he displayed great facility in the mental handling of numbers, multiplying readily and correctly two figures by two, but failing in attempting numbers of three figures. My tutor, a Cambridge man, Fellow of his college, strongly recommended the father not to carry his son about the country, but to have him properly trained at school. This advice was not taken, for about two years after he was brought by his father to Cambridge, and his faculty of mental calculation tested by several able mathematical men. I was present at the examination, and began it with a sum in simple addition, two rows, with twelve figures in each row. The boy gave the correct answer immediately. Various questions, then, of considerable difficulty, involving large numbers, were proposed to him, all of which he answered promptly and accurately. These must have occupied more than an hour. There wa

subsequently one of the great English lords of Norway, with a very eligible interest in that snug little railway and the Miosen navigation. Not far from the western extremity of the Miosen lake is the prettily situated village of Lillehammer. From Lillehammer the tourist may proceed northwards to Trondhjem, or westwards to Bergen and Hardanger, or through the Guldbransdal to Dovrefjeld. This is described by Williams as a vast undulating moorland,

---

then a pause. To test his memory, I then said to him, " Do you remember the sum in addition I gave you?" To my great surprise he repeated the twenty-four figures with only one or two mistakes. It is evident, therefore, that in the course of two years his powers of memory and calculation must have been gradually developed. He could not explain the process by which he worked out long and intricate sums. He did not appear at all overworked. As soon as a question was answered he amused himself with whipping a top round the room, and when the examination was over he said to us, " You have been trying to puzzle me, I will try to puzzle you. A man found thirteen cats in his garden. He got out his gun, fired at them, and killed seven. How many were left?" "Six," was the answer. " Wrong," he said,—" none were left. The rest ran away." I mention this to show that he was a cheerful and playful boy when he was about ten years old, and that his brain was not overtaxed.'

Such powers in a boy supply no indication that he will be distinguished in after life. Many have lost the trick by which the calculations are made while they are yet young, most have proved in after life nothing beyond more of average power, and none of them, it is said, have exhibited the slightest tincture of genius in mature life.

Bidder, after being for a time a world's wonder, had the good sense to study carefully to qualify himself for the profession in which he engaged; he became a civil engineer of some eminence; he enjoyed the confidence and esteem of Robert Stephenson; and rose to be president of the Institute of Civil Engineers. He has been dead for some years.

In further illustration of what has been done in mental arithmetic, I cite the following extracts from letters from Dr John Wallis, of Oxford, to Mr Thomas Smith, B.D., Fellow of Magdalene College, which are preserved in the *Classical Journal*, vol. xi., No. 21, p. 179, and were republished in the *Spectator* of January 4, 1879 :—

'December 22, 1669.—In a dark night, in bed, without pen, ink, or paper, or anything equivalent, I did by memory extract the square root of 30,000, 00000, 00000, 00000, 00000, 00000, 00000, 00000, of which I found to be—1,77205, 08075, 68077, 29353, *feré*, and did the next day commit it to writing.'

' February 18, 1670.—Joannes Georgius Pelshower (*Regiomontanus Borussus*), giving me a visit, and desiring an example of the like, I did that night propose to myself in the dark, without help to my memory, a number in 53 places, 24681357910121411131516 1320171921222426283032325272931, of which I extracted the square root in 27 places—15710301687148280581715217l *proximé*, which numbers I did not commit to paper till he gave me another visit, March following, when I did from memory dictate them to him. Yours, &c.'

In Bayle Dr Wallis is described as a man of very great attainments, with a peculiar character for deciphering, and altogether very successful in life. The correspondent of the *Spectator* remarks :—'The *naïveté* of the *feré* and *proximé* is charming, and also the confession that he did not commit the appalling row of figures to paper, but dictated them a month afterwards to his friend from memory. These feats are perhaps not so difficult as multiplying 15 figures by 15, for while of course it is easy to remember such a number as three thousand billion trillion being nothing but noughts, so also it may be noticed that there is a certain order in the row of 53 figures ; the numbers follow each other in little sets of arithmetical progression—(2, 4, 6, 8), (1, 3, 5, 7, 9), (10, 12, 14), (11, 13, 15), (16, 18, 20), and so on ; not regularly, but still enough so to render it an immense assistance to a man engaged in a mental calculation.'

between three and four thousand feet above the sea level. It has no particular claims to the picturesque, and the absence of great rocky masses deprives it of any savage grandeur though it is sufficiently desolate. The tints of the abundant reindeer moss, or rather lichen (*Cenomyce rangiferina*), are in many parts very beautiful; especially where a rounded heap of earth-covered boulders is overgrown with it. It is dry and crisp, forming a luxurious mountain couch; it varies from straw-colour, through a pale buff, to a bright orange and red brown. Its habit is to grow on the dry, well-drained spots, while peat moss occupies the swampy localities.*

Du Challu, in his work entitled *The Land of the Midnight Sun*, has given a graphic sketch of reindeer grubbing under the snow for this article of their food; and he has given a genial account of the domestic life of the farmers in Guldbrandsal.

In continuation of his journey, after describing Jerkin and the hospitalities of the hostelry there, Williams says:— 'Walking on the fjeld the view of Schneehaettan is rather fine from its highest ridge. This mountain, long regarded the highest in Norway, is not so imposing as might be expected from its height, 7,620 feet above the sea; but it is only 4,500 feet above Jerkin, and 3,520 above this point, which is 4,100 feet above the sea level, and said to be the highest carriage road in North Europe. The ascent appears very easy from here—a long ridge stretching gradually down from the summit like Goat Fell in

---

* The botanic name of the lichen, *Cenomyce*, is derived from *kenos*, empty, and *mykos*, a minute fungus, and has been given in allusion to the hollowness of the little fungus receptacles with which it is studded. This constitutes for the greater part of the year, and especially in winter, the food of the vast herds of reindeer on which the Laplanders further to the north are dependent for support—hence comes its popular name. Linnæus tells that no vegetable grows throughout Lapland in such abundance as this, especially in woods of scattered pines, where for very many miles together the surface of the sterile soil is covered with it as with snow. On the destruction of forests by fire, when no other plant finds nutriment, this lichen springs up and flourishes, and after a few years attains its greatest size. Here the reindeer are pastured, and whatever may be the depth of snow during the long winters of that climate, they have the power of penetrating it and obtaining their necessary food. Linnæus has given a beautiful description of this lichen, and of the animals whose support it is, in the *Flora Lapponica* (p. 332). It may be found in woods in Britain, and another species found growing on banks—the Cupped conomyce, *C. pyxidata*—is sometimes employed by the poor in the cure of the hooping-cough.—*Louden.*

Arran. But appearances from such a distance are little to be relied on, especially about the region of the snow line. Professor Forbes, who is no novice in mountaineering, describes the ascent as very troublesome, on account of the deep sludgy snow-pits between the boulders.

'The road now plunges into a deep valley in company with the river Driv, which roars and foams among the rocky masses that restrain its course. The amount of water at this elevation gives evidence of the extent of the fjeld, and of the quantity of snow that is thawing around Schneehaettan. Many small lateral streams pour into the valley, cutting deep gutters in the rocks over which they fall. Several of these flow directly from the patches of snow that fill the hollows above. There is a curious and very pretty effect produced by a peculiar formation of the mountains on the other side of the river. Each ridge of rock runs down nearly parallel with the valley, and forms a long slender-pointed high-backed promontory; one side of the promontory ridge being nearly perpendicular, and thus a little blind glen is formed into which the rocky promontory would about fit if it were reversed. These glens are evenly curved, and smooth, covered with rich grass, and dotted with shrubs and lilliputian birch trees. They are very numerous, much alike, and occur at rather regular intervals, giving quite a character to the valley, and contrasting beautifully with its general wildness: any one of them would form a subject for a charming little picture. The scenery is very fine all down this ravine to Drivstuen. The river makes some fine cascades, and several minor ones are formed by the streamlets which tumble into it from the snow patches.

The character of the scenery changes below Drivstuen, where, instead of wild broken rocks, the road passes over an almost park-like green sward. . . . My next day's walk was through a rich cultivated valley, with snowy mountain peaks ahead. Murray says the Schneehaettan is visible here, but unless my maps and compass deceive me considerably this must be a mistake.

'A little before reaching Ovne, or Aune station, there were some of the most magnificent banks of pansies I ever beheld. Several patches of above a hundred square yards were covered with an unbroken carpet of these beautiful little flowers, the variety, richness, and harmony of their colours were most exquisite; they saturated the atmosphere around with a delicious aroma, which was almost intoxicating in its concentration. I lay down upon them and slept for an hour or two, the sunbeams poured upon me with a roasting heat, the rooks were cawing above, and the river tumbling below, though yesterday and this morning it was freezing, and the snow patches were still visible in all the hollows of the craggy rocks above. I dreamed of Oriental vapour baths, otto of roses, and beautiful primroses just imported from the snowy Caucasus, and selling in Covent Garden for a few shillings per dozen.'

Schneehaettan is visible near Staen, about fifteen miles below the place where Murray speaks of it. It is a more picturesque object from this point than from the Dovre fjeld. A number of other snowy peaks are also visible. Lakes are pretty numerous in Norway, but they are comparatively small. While in Sweden there are 34 lakes of more than 100 square kilometres in size, in Norway there are only 52 which exceed 25 square kilometres. The superfices of all the lakes of Norway is 7600 square kilometres, or 2·4 per cent. of the whole area of the country, while in Sweden lakes constitute 10 per cent. of the area of the country. They abound most in Southern Norway. There, in eight prefectures, is found one-half of the superficies of the lakes of the whole country; and they constitute 4 per cent. of the area of these prefectures. Many of the Norwegian lakes are of considerable depth, so much so, that over wide districts the bottom is below the level of the sea. Sometimes even they are deeper than are the greater adjacent fiords. Thus the great lake of Norway—the Miosen—has a depth of 457 metres, even in the low water of winter. Its surface is then only 121 metres above the level of the sea, consequently its basin, over a great extent, is 330 metres (1110 feet) below the level of the sea, which only attains to this depth in the outer portion of the Christiania fiord.

The Storsjo, in the valley of Rendal, has a length of 35 kilometres, and an area of 47 square kilometres. It empties its waters into the Glommen. Its altitude is 257 metres, its depth is 301 metres, its basin sinks 44 metres (147 feet) below the level of the sea, from which, by the course of the Renelv and the Glommen, it is 340 kilometres distant.

The Tyri fiord, with an altitude of 64 metres, has a depth of 281 metres, so that the bottom is 217 metres (723 feet) below the level of the sea.

Many small lakes situated near the interior extremity of the deep fjords of the prefecture of Bergen and Romsdal have a depth as great as that of the fiords to which they are adjacent. Thus the Horningdalsvand in the prefecture of Romsdal, at an altitude of 56 metres, with an area of 57 square kilometres, has a depth of 486 metres, so its bottom is 429 metres, 1,430 feet below the level of the sea, from which it is distant only 10 kilometres. On the south-west coast, in the low lying level country of Jæderen, there are lakes of fresh water, separated from the sea by only banks of sand, with which they have nearly the same level. The largest of them, the Orrevand, measures 11 square kilometres.

It may be premature to speak of the means by which valleys of such depth have been dug, or ploughed out of the solid earth. But the subject will not be overlooked; it will be discussed in the chapter entitled *Mechanical Action of Glaciers*. Here it is deemed sufficient to intimate that these are considered as valleys occurring in the *thalweg* or bed of a river or a rivulet, which was filled before the stream could continue its course beyond, and which are kept full by supplies equivalent to some extent to the current or delivery passing on.

A tabulated list of eighty rivers and main water-courses of Norway, their names and length, the superficial area of the fluvial basin, and the debouchure of the same, and a like tabulated list of about a hundred and fifty of the principal lakes in Norway, their names, locality, superficial area, length, and altitude, are given by Dr Broch in his report on these.

CHAPTER XI.

WINDS.

An important and manifold influence on the distribution of forests, and of different kinds of forest trees, is exercised by the winds. By means of these seeds of trees are widely dispersed; they are thus borne to bare spots in the forest, and to bare lands beyond; and the winds thus contribute where local conditions are favourable, to give extension to the distribution of forests, and of different kinds of forest trees. But frequently their operation is also destructive. By successively stripping a tree, young or old, of its foliage, the tree is starved and it dies; by bringing breezes from the sea charged with salt they kill many kinds of trees, and thus prevent the growth of such along the coast; by breaking off a bough or a twig they give access to moisture, and to germs of fungi, either of which may prove fatal to the tree; by blowing vapours from a kiln they may occasion the death of an extensive forest patch; by spreading a forest fire they may devastate a whole country side; and by a gale or cyclone they may lay low trees the growth of centuries.

But not less important, on the other hand, is the action of forests as a windbrake giving shelter to dwellings and cultivated fields; and in arresting malaria; and in preventing the drifting of loose sand, which otherwise might render sterile fertile lands. In view of such effects the conservation and extension of forests has in many countries been made subject of legislation and government administration in which the rural economist is not less directly interested than is the student of forest science.

In some countries to the south of Norway much attention is given, both in forest legislation and in forest administration, to the conservation of forests on mountain tops as a means of giving shelter to lower-lying lands to leeward, and as a means of preventing the formation of mountain torrents. I have no reason to suppose that the subject commands like attention in Norway. But meteorological observations of storms and prevalent winds, and of atmospheric pressure are not awanting, and in view of the relation of winds to forests, I cite some of these reported by Dr Broch.

Thunder storms and storms of rain are comparatively rare in Norway. They principally occur in the summer months. The winter storms fall almost exclusively on the west coast, from the Naze to Andenaes, in Vestrealen; they come in heavy gales from the west.

The thunder storms come as a general rule with wind from the south and the south-west, of a high temperature, and loaded with aqueous vapour. They proceed often in a regular course across the country, and sometimes advance with a great extent of breadth, for example, from the Naze to Vest fiord. Their mean rate of advance is 40 kilometres per hour, in a direction from south-west to north-east. The greater part of them are accompanied by great whirlwinds along the south border. A portion of the storms are occasioned by the great heats which are produced in the interior of the country. These are amongst the least violent.

The storms of Norway are everywhere greater and more frequent on the coast than in the interior. In the eastern part of the country, where they only appear in summer, there may be two or perhaps three in a year. They are also of like frequency in the interior, especially on the north part of Lake Miosen, as they are on the coast. It is on the shore, from the Naze to the Sogne fiord, that they are of greatest frequency in Norway; six or seven have been seen in the course of a year. From the Sogne fiord to the Drontheim fiord they are comparatively rare. They are more frequent in the region which extends from the Drontheim fiord to the Arctic Circle. To the north of the Arctic Circle their frequency diminishes considerably; and on the coast of Finmark thunder is never heard excepting in very hot summers. In South Varanger, on the contrary, the summer storms are more frequent, and there thunder is heard as often as in Southern Norway.

Lightning rarely strikes the ground, still it causes many accidents in the course of years, especially in winter thunder storms; and on the coasts of Southern Norway many churches have been struck by lightning.

The occurrence of thunder storms has a marked diurnal time during summer. The greater number are in the afternoon, and there are fewest during the night. During winter the nocturnal storms seem to be as frequent as those which occur during the day. The hail accompanying thunder storms in Norway is in quantity insignificant.

Next in importance to thunder storms and storms of rain, are storms of wind. These storms attain their greatest frequency in Norway in the month of November, during which there may be 4·5 days of storm counted on. They are least frequent in summer, when they average ·8 days per month. This frequency increases in autumn and decreases in spring in a regular way. There is no marked frequency of storms during or immediately after the equinoxes.

The frequency of storms and the force of the wind follow the same law. The frequency attains its maximum on the coasts of Nordland and of Finmark, where there are on an average 46 in the year; the west coast follows with an average of 27 per annum; their number diminishes on the coast of the Skagerack, where there occur about 15 in the course of the year; and lastly, in the interior they are comparatively rare, numbering in Christiania from 2 to 3, and on the Dovre mountains from 5 to 6 a year.

The direction followed by them is generally that of the dominant winds: here let it suffice to state that from Christiania to Vardo, coming generally in winter, they come almost always from the south-west. On the west coast they blow most frequently from the south or the south-west, but also pretty frequently from the west or from the north-west. At the Naze storms from the west predominate, and on the coast of the Skagerack those from the south and south-west.

Storms from the south-west are the most prevalent over the whole land; and next to these are those from the west. Storms from the east are the most rare; and storms at sea are much more frequent than storms on the land.

In peculiar circumstances, as when in winter the cold of the continent is very intense, and the contrast between it and that of the sea very great, there may spring up very strong land winds. These come down, especially in Southern Norway, where the climate of the interior and that of the coast come pretty near, and where the current of air seeks the fiords, which set free a heavy cold air; this cold air precipitates the warmer sea vapours, which spread themselves over the water in a thick cold mist. It is in these circumstances that the greatest degree of cold is experienced on the coast, although pretty often during a time of calm the temperature of the interior comes down much lower.

From winter to summer the direction and force of the usual winds so varies that the information to be supplied may eludicate but little. On the west coast and in Southern Norway, from Skudesnaes to Vardo, and on the the embouchure of the Christiania fiord, winds from the south and south-west predominates. At Lister and at the Naze west winds, and at Christiania winds from the north, prevail. For the whole country it is the south-west wind which is the predominant wind; next to these are winds from the south; the winds least frequent in Norway are those from the east.

In winter the predominant winds are—from Christiania to the Naze, the nort-east; at Lister, the east; along the west coast, the south; from Cape Stat to Drontheim fiord the south-west and the south-east; on the Dovrefjeld, the south; at Folden, the south-east; at the Lofoden islands, the west-south-west; at Andenæs, the south-south-east; at Vardo, the south-west. In other words, the dominant winds follow the direction of the coast, keeping the land on the right hand.

The tendency of the winds to go along the course is so strong that in both seasons there are twice as many winds blowing along the shore as against it. This preponderating influence on the coast seems to lose itself pretty rapidly as the landward distance from the shore increases; and tabulated observations at Utsire, 59° 18' N.E. Geenwich, and at the outer-lying Lofoden islands, show that the winds there blow so equably in all directions that none have a marked preponderance. In other stations, on the contrary, it is easy to indicate the normal direction of the wind.

The force of the wind is much more considerable on the coast than in the interior of the country. On the coasts of Finmark, at the Lofoden islands, and at Utsire— that is to say, at the places most exposed to sea winds, the force of the wind has its maximum. The mean force of the wind there amounts to that of a pretty strong breeze; in the interior of the country, on the contrary, it scarcely amounts to that of a very gentle wind. At Bergen, and on the coast of the Skagerack, the wind has the force of a moderate breeze. At Skudesnaes the wind blows with a mean force three times that experienced at Christainia.

It is in winter that winds on the coast attain their greatest force, and that calms are least frequent, while during that season the feeblest winds and calms prevail in the interior of the country. In summer it is exactly the reverse. At this season calms on the coast are most frequent, and the winds are weakest; while in the interior, provided always the distance from the sea be not too great, the winds are strongest and calms the most rare. The difference in the force of the wind in summer and in winter is much greater on the coast than in the interior of the country.

The diurnal variation in the force of the wind is very little marked in winter; it is more pronounced in summer, and especially in the eastern portion of the country is this the case. The force of the wind attains its maximum about 2 P.M.; it sinks to its minimum during the night.

Thus there is on the coasts of Norway a perfect contrast between the direction of the winds in summer and in winter. The chief occasion of this must be sought in the alternate distribution of the pressure of the air on the central portion of the Scandinavian peninsula. In winter the air experiences a condensation which causes an onward current of winds issuing from it; and in summer a rarifaction which attracts a current of air or of wind towards the locality.

For all this information I am indebted to Dr Broch's report.

CHAPTER XII.

GEOLOGICAL FORMATIONS.

THAT the geographical distribution of vegetables is greatly affected by difference of soil in different localities is patent to all; and that these differences are attributable largely to underlying geological formations may be readily admitted.

There are plants, such as the lousewort, the butterwort, and the marsh marigold, which we never find growing excepting on marshy spots; there are others, such as the ling and the heather, which we find only on what are known as heath-lands, which consist largely of peat soil; there are others, such as the coltsfoot, found only on clay land; others like the oyster plant, only on sand; and others, as are most of those with which we are conversant, are found only on what is called garden soil. And agriculturists, with all the appliances of modern science at their command, find it not only more profitable, but almost necessary, if they would secure remuneration for their outlay of money and labour, to confine themselves to the culture of certain crops on light lands, and of others on heavy lands, and it may be of others still on varieties of these as numerous as the notes of the gamut.

A carefully conducted analysis of any plant by burning, and by a chemical analysis of the ashes, may be made to show that there were in the plant besides carbon and moisture, which may have been obtained in one way of another from the atmosphere, mineral substances which must have been obtained from the soil; and an analysis of the soil, made before the growth of the plant and after, may be made to show that what the plant got the soil lost. From this it may be inferred that unless what has been thus withdrawn, or its equivalent, be restored by the decomposition of rocks allowed to go on while the land lies fallow or is utilised by the culture of some other kind of crop, or by a supply of it in manure, the soil will sooner or later become exhausted of these constituents; and observation has shown that then the land will no longer yield that crop. So is it with land altogether devoid of constituents required in the production of the same or other vegetables. And thus is the geographical distribution of trees, and of arborescent as well as of herbaceous plants, to a great extent determined.

It may be instructive to bear in mind that what is required for the growth and reproduction of a plant, the seed of which has found its way to any spot, is not only heat and moisture, but these in quantity, defined within a limited range, varying in many cases with the successive stages of germination, growth, flowering, ripening of the fruit, and the casting of the seed, a departure from which at any one of these stages, or at some stages intermediate between them, imperilling the result. But not less essential and necessary to the continued growth of the plant is a soil supplying in appropriate form and in appropriate quantities the nutriment required by the plant.

'The beautiful indigenous plant, the "lady's slipper," grows,' says Schleiden, 'over all parts of the Swiss Fore Alps, where the soil is formed of the Alpine limestone; it accompanies the whole Swabian musselkalk, and disappears suddenly when we come to the sand of the Jura and keuper formations on this side of the Danube. It next makes its appearance on the musselkalk of Turingia, and comes down with that on the Werr as far as the neighbourhood of Gotlengen, and it leaps over the Bunter sandstone of the lower Eichsfeld, and the granite o the Upper Hartz, and it again gladdens the eye of the wanderer on the calcareous formations eastward of the Brocken.

'It is sought in vain over all the clay and sand formations of the northern German plains, till, in the extreme north, it again shows itself in Rugen, where the chalk rocks of Arkend and Stubbenkammer lift their heads.

'Again, on the western coast of France, there grow various insignificant-looking shore plants, species of the *Salsola* and *Salicornia*, which the inhabitants use to obtain soda from the ashes. When we travel from thence toward the east, we everywhere miss these little plants, even when searching most carefully, and merely one or other of them makes its appearance in places where the soil is moistened by some salt spring. At last we arrive at the great steppes of the south-east of Russia, which in summer are covered with a thick crust of salt, and here these plants are found growing in the same abundance and luxuriance as in the west of France.

'On the northern coast of Germany the little pale maiden pink grows upon the arid sand downs, and is universally distributed over the sandy plains of northern Germany; but these are succeeded by the granite, clay, slate, and gypsum of the Hartz, the porphyry, and the musselkalk of Turingia, and our little pink is not met with again till we arrive on the keuper sand plains, on the further side of the Maine, surrounding the venerable city of Nuremberg. It extends further south through the Palatinate, till the musselkalk of the Swabian Alps again sets a limit to it, but it leaps over these and the whole Alpine region, and at last appears again on the sandy soils of Northern Italy. How is it these plants everywhere disdain the richest soils in their range of geographical distribution, and are conformed to perfectly determinate geognostic formations? Must not the lime, the salt, the sand, or rather the *Silex*, have a most distinct influence in the matter?

'In regard to soil science has gone astray in the most varied and opposite directions. So late as the commencement of the present century there were men who asserted that plants could themselves form all their organic and inorganic constituents out of the air and distilled water; and there were superficial experiments which seemed to countenance the idea. Subsequently the error lay in the other extreme, for there was manifested a disposition to ascribe a peculiar flora to each geognostic formation.

'The truth seems to lie between the two extremes. When we find that the ashes of tobacco, of clover, of lucerne, contain more than 20 per cent. of lime and magnesia salts, we cannot be surprised if we do not meet with them on pure sandy soils containing scarcely a trace of lime; but it would be drawing a false conclusion from this to say that the musselkalk, or the keuper limestone, or the Jura limestone, or any other calcareous stratum of any given formation is exactly the proper soil for these plants.

'That a plant like the great sugar tangle (*Laminaria saccharina*), which is so rich in soda, iodine, and bromine, occurs only in the sea, and not in fresh water, where soda is very sparingly, and iodine and bromine not at all present, is certainly easily conceivable. But it is certain, at the same time, when we decide upon the soils on a large scale, according to geognostic principles, that there are very few plants characteristic of particular constituents, and this relation is, indeed, neither very natural nor necessary. In the next place it may be asserted that all plants contain the same constituents in their ashes, but in very different proportions.

'On a soil, therefore, composed purely of one kind of earth—*e.g.*, lime, silex, or gypsum—no plant at all could flourish. Every soil that bears plants contains also in its composition all the substances required by all plants, only the proportions differ, and the predominance of silex, lime, or common salt, must consequently favour especially the growth of grasses, pulses, or other plants, although these are by no means exclusively confined to the proper sandy or calcareous soils, or to the sea-side. In reference to this point,' says Schleiden, 'I know really no other plants than the carbonate of lime plants, the gypsum and salt plants, which I could bring forward in evidence.

GEOLOGICAL FORMATIONS.

'The mechanical condition and physical peculiarities of the soil will also be found to modify the effects of the influential circumstance already referred to, and contribute to chain particular plants to particular soils, or to facilitate their dispersion.

'There are plants which will only settle on unbroken rocks, and which, when the other conditions coincide, spring from these rocks to our walls, like the wall rue spleenwort (*Asplenium Ruta muraria*), a little fern, the name of which denotes its station. Others occur only where weathering has broken up the solid rock into small fragments—*drift* plants, which, clinging to mankind, select rubbish heaps, which most resemble their natural station: the great nettle and the henbane serve as examples.

'Lastly, other plants grow only where the rocks have been reduced to fine powder in sand, or in the fine-grained clay produced by chemical decomposition. The so-called German sarsaparilla, the sea weed (*Ammophila arenaria*), is an example of the first condition, but there is no definite condition corresponding to it in the vicinity of human habitations.

'Clay, on the other hand, stands beside the black substance, *humus*, resulting from the decomposition of organic matter. Both rich in soluble salts important to vegetation, both distinguished in their property of absorbing from the atmosphere, and their conveying to the root of plants, gases and aqueous vapour, they cause, single or in combination, the most luxuriant vegetation. And we thus obtain three stages in reference to the qualities of the soil —pure earths, wholly devoid of vegetation; mixed earths, without clay or humus, with an arid but characteristic vegetation; and lastly, soil rich in clay and humus, with the greatest abundance and variety of plants.

'Even in the north the eye of the uninstructed observer is struck by the greater richness and stronger development of the vegetable kingdom, from the argillaceous, basaltic, and porphyritic soils; and under the tropical sun, simple quartz sand is a desert, if water, and therein foreign matters, be not furnished to it. Witness the Great Sahara and the Sahara of the South.'

In view of such observations as these made elsewhere, the student of forest science, in studying phenomena presenting themselves for consideration in Norway, may be disposed to give his attention for a little to the geology of the country, and to the distribution of trees in relation to this. For the following sketches of both, I am indebted to the report by Dr Broch. It is only because of the freeness with which I happen to have abbreviated and translated his statement that I do not mark it as translation.

The fundamental rock which consist of gneiss in primitive schists—mica schist, talc schist, pot stone, quartzose schist, quartzite, &c., appears on the coast of the Miosen lake and of the Christiania fiord, from Elveram to the south of the plateau of the Oplande, and in continuation to the south. It is traversed by small eruptive masses of old granites, partly in chains, partly in masses, which in the Solor and the Odal take all the direction from N.N.W. to S.S.E. The frontier region of the Odal, the Finskog, and the wooded ridges of the Vinger, consist of eruptions of this kind, and especially of striated granite. These eruptions form wooded ridges of little elevation, and often also of little breadth. Elsewhere they are met with in various districts, and are generally wooded. They are often accompanied by *gabbro*, consisting of hyperstene, labradorite, and nerite; and metallic ores or metallic veins and deposits are not awanting.

Elsewhere there are met with masses of granite of later eruption seen in islands lying outside of the Drontheim fiord, and elsewhere. And in some adjacent islands may be seen the fundamental rock covered with an envelope of sandstone grit and conglomerate.

'The granites, which cover an area of 23,000 square kilometres, inclose, especially in the depth of valleys, parts more or less distinctly marked, kinds of islands where the fundamental rock makes its appearance. And in these, as in all the eruption rocks, there are often seen in the granite mass quite small fragments, with sharp edges of this fundamental rock. The granites of this great rock collection are partly granulated, partly striated. The old granite presents an appearance by no means attractive, especially on the mountains of the Sætersdal. It forms there bare rugged cliffs and fields sown only with blocks of stone, and here and there covered with with marshes. In the lower lying countries, especially to the east, better sheltered from the sea winds, this granite supplies, on the contrary, an excellent ground for forests.

Immediately after the great upturnings of the ground which accompanied the eruptions of the old granite, or perhaps while these eruptions were still occurring, there was produced the *taconic* formation, composed of three distinct strata or layers, resting in discordant non-conformable stratification on the fundamental rock. The lower of the layers, particularly distinguished by the quantity of fragmentary rock contained in it, has been called the *Sparagmitic* layer from the name given to one of these rocks. *Sparagmite* is a pudding stone or conglomerate, sometimes reddish, sometimes grey, in colour, consisting of fragments of felspar and quartz, often intermixed with broken talcose scales.

Where the quartz prevails, and is not covered with a vegetable soil, the ground is arid, as on both banks of the Glommen, from the Marafjeld to the valley of the Lilleelvdal. In districts sheltered from the wind, however, the *Sparagmite* offers a ground pretty favourable for forest vegetation.

The stratification of it is disposed in some places in inclined beds, or on plateaux in horizontal layers. On the Rundane these latter are cut up into pyramidal mountains, the sides of which make apparent the horizontal layers.

The great quantities of fragments of felspar and quartz seem to be the result of the decomposition of the old granite. The mica is found in like broken state in the argilaceous schists of this layer. Above, on its surface, the layer presents on many parts a hard conglomerate, erratic blocks of quartz in which appear to have come from ancient quartzose mountains in like manner destroyed, and among these blocks are also found fragments of the old granite, pure and simple.

No fossils have been found in this layer; but it is considered as the base of the *taconic* formation, and it lies in a position divergent from that of the horizontal strata on vertical layers of the fundamental rock. Fossils have been found at Hogberg, near the river Klara Elv, to the south of Kvitvolfjeld.

Sandstone is met with on the surface of the sparagmitic land in the part to the east of Trysel, whence are obtained the famous white stones bearing that name.

On the sparagmitic formations in the western part of the plateau of the Oplande rests the second layer of the *taconic* formation, composed of argilaceous micaceous schist and limestones, with rarely occurring *débris* of nascent animal organisms. From the most distinctly characteristic portion of this layer composed of black argilaceous schist containing petrifractions of dictyonémes—a species of bryozoes; and from species of the genus of trilobytes called olenus being found in the limestone, this layer has received the designation dictyonema schists and olenus limestone. It corresponds to the period of the Potsdam in the United States. It covers an area of 4,500 square kilometres. The argilaceous schist is readily disentigrated, and produces excellent pasture land on the mountain summits, and in many valleys a rich arable land. The same may almost be said of the limestone, the fertility of which, from the decomposition of animal remains, is considerable.

In some places, as on the west shore of the Christiania fiord, at the Skien fiord, to the south of Kongsberg, and both banks of the south part of the Miosen lake, the layer of argilaceous schist lies immediately on the fundamental rock.

Above this layer, in some mountainous region, is found the third, known as high mountain quartz, with transition schists, and beds of dolomite and gneissic schists. The high mountain quartz is sterile and poor.

These layers are in many places traversed by great subsequent eruptions of gabbro [basalt?], of which are formed a multitude of mountain summits and plateaux. The rock thus produced is composed mainly of granite, syenite, and greenstone, and metaliferous minerals are found collected in many places on the confines of this rock. On the Dovrefjeld are found more recent eruptions of trap.

The schistose and limestone region, along the limits of the wall of exterior granite, manifests transformation in a way perfectly evident. Their broken or inclined beds seem metamorphosed, each according to its primitive nature, in a manner analogous to what may be seen in the comparatively much more accessible and more explored silurian region of the Christiania fiord. The calcareous stones have crystallised into marble, the argillaceous schists and the sandstones have become schists of micaceous gneiss. Garnet is often found developed in the schists; and at the Kjæringfjeld, to the south of the Sorfolden fiord, the micaceous schist encloses beryl.

On the layers of the taconic period, which have been described, there rest in Southern Norway silurian formations disposed in many layers, and finally layers of sandstone and conglomerate, corresponding to the Devonian formations.

In some places the second layer of the taconic formation —the dicytonema schists and olenus limestone—supports thick layers of gray, impure limestone, alternating at times with dark, argilaceous schist. The limestone encloses a quantity of fossil trilobites, especially species of the genus *Asaphe, Orthoceras vaginatum*, belonging to the cephalopods, and *graphtolites*, a species of polypes. From these fossils it has received the name of ancient graphtolite schist and *Orthoceras vaginatum* limestone. Above this is a second layer designated the trinucleus and chasmoss limestone, and a third called the layer of gray limestone. These are all comprised in the lower silurian formations, and above these are three superior silurian formations belonging to a later period.

Rarely are all these layers found superimposed in a complete series. Ordinarily there are found some parts of the series exposed in steep stratifications, which, being traced, are seen to belong to great corrugations which have been produced by a powerful compression of the whole system of layers, by means of a stupendous force accompanying the appearance of the great later irruptive masses. Subsequently, the projecting portions of the corrugations having been worn away, their remains appear as dislocated stratifications.

Above the upper silurian layers, with their abundant fossils, are great beds of red argilaceous schists, red and gray coloured sandstones, and conglomerate, supposed to belong to the Devonian formation or old red sandstone; but no trace of fossils has yet been found in them. Mention must also be made of a field of sandstone of uncertain age, but apparently still older.

Subsequently to the period of the first appearance of swarming animal organisms in the middle of the silurian period, and repeated perhaps long after the Devonian epoch, we see that anew great eruptive masses of different kinds have been thrown up from below, sometimes in lines straight as if drawn by a cord, sometimes in small isolated spots, sometimes over great areas, and sometimes as a stream flowing over the earlier formed layers.

Some of the granite of this region is, as is the case in the Grefsenaes, near Christiania, very easily wrought in long comparatively narrow pieces; and being susceptible of a very fine polish, it is exploited on a great scale for all kinds of building purposes: stairs, kerbstones, sockets, pillars, tombstones, &c.

In the time of these eruptions the ground was also broken up by long fissures which, opening from below, were filled with granite, porphyry, and greenstone. In these fissures all kinds of fragments are met with, and many of these dykes extend without alteration beyond the layers of schist, limestone, and sandstone, which they traverse, on the one hand, into the great eruptive masses of granite, syenite, and porphyry, and on the other hand, far beyond the limits of the silurian region into the beds of the fundamental rock. These dykes inter-cross one another regularly; the greenstone dykes are thus seen to be notably the latest formations.

In Southern Norway the series of formations cannot be traced beyond the commencement of the Devonian period. Covering these immediately are those of the glacial period. All the intermediate formations found elsewhere in Europe, such as those of the coal measures, and the secondary and tertiary formations, are entirely awanting.

In the north there is found in the island of Ando, the most northern of the islands of Vesteraolen, a sandstone formation with intercallated beds of coal, and of combustible schists of little importance. They come to the surface. The fundamental rock seems to take there the form of a cup, which is filled with beds of sandstone, and is now covered with an extensive marsh. Elsewhere there are sandstones and conglomerate of uncertain age, but supposed to be Devonian. In the Beskadesfjeld are found beds of graphite supposed to belong to the carboniferous period. In the peninsula of Varjag, all the heights are covered with a reddish brown conglomerate and brown sandstone, supposed to belong to the Permian period; and it may be remarked in passing, that all the larger rivers which traverse the plateau of Finmark are auriforous. Gold is found in small leaves or scales in the beds of the rivers, and in the large gravel of the erratic deposits through which they flow. The formation of these deposits is attributable to what is known as glacial action, at a later period of the earth's history, which will be brought under consideration in connection with other phenomena which are attributable to this.

I

CHAPTER XIII.

MOUNTAINS AND FJELDS.

It may have been observed that in the details given in regard to the geographical distribution of different kinds of forest trees in Norway, mention is made chiefly of latitude, but the fact that such distribution is affected by altitude as well as by latitude must not be ignored. There is a co-relation between the two in so far as temperature is concerned; and it is mainly in relation to temperature that the relation of altitude in different localities at which different trees grow becomes of importance in discussions connected with this fact. But the difference observable is not related to this exclusively. The soil is generally different in the plain, and even at the base of a mountain, from what it is on the sides, and on the summit of that mountain; and there is also a difference in the atmospheric pressure at a high altitude, and at the level of the sea, and at all intermediate altitudes, which difference of pressure may not be without effect on the geographical distribution of trees.

To the establishment and acceptance of this as a fact, it is not necessary that we should know how it is that it produces the effect; it is a fact determined by observation; and it may be seen on the ocean as well as on dry land.

'The ocean, as well as the land,' writes Balfour, 'possesses its vegetable forms, which are of a peculiar kind, and exist under different conditions of pressure, of surrounding medium, and of light. Some sea-weeds, Harvey remarks, are cosmopolitan or pelagic, as species of ulva and enteromorpha, which are equally abundant in high northern and southern latitudes, as they are under the equator and in temperate regions. Codium tomentosum, ceramium rubrum, C. diaphanum, species of ectocarpus, and several conferæ, have a region nearly as wide. Plocamium coccineum and gelidium corneum are common to the Atlantic and Pacific oceans; rhodymenia palmata, the common dulse of Britain, is found at the Falkland Islands and Tasmania. Fucus tuberculatus extends from Ireland to the Cape of Good Hope; fucus vesiculosus occurs on the north-west coasts of America, and on the shores of Europe; while desmarestia ligulata is found in the north Atlantic and Pacific oceans, as well as at the Cape of Good Hope and Cape Horn.

'In general, however, sea-weeds are more or less limited in their distribution, so that different marine floras exist in various parts of the ocean. The northern ocean, from the pole to the 40th degree, the sea of the Antilles, the eastern coasts of South America, those of New Holland, the Indian Archipelago, the Mediterranean, the Red Sea, the Chinese and Japanese seas, all present so many large marine regions, each of which possess a peculiar vegetation. The degree of exposure to light, and the greater or less motion of the waves, are very important in the distribution of algæ. The intervention of great depths of the ocean has a similar influence on sea-plants as high mountains have on land-plants. Laminariæ are confined to the colder regions of the sea; sargassa only vegetate where the mean temperature is considerable. Under the influence of the Gulf Stream, sargassum is found along the east coast of America, as far as lat. $44°$; and the cold south polar current influences the marine vegetation of the coasts of Chili and Peru, where we meet with species of lessonia, macrocystis, d'urvillæa, and iridæa, which are characteristic of the antarctic flora. Melanospermeæ, according to Harvey, increase as we approach the tropics, where the maximum of the species, though, perhaps, not of individuals, is found; rhodospermeæ chiefly abound in the temperate zone; while chlorospermeæ form the majority of the vegetation of the polar seas, and are particularly abundant in the colder temperate zone. The green colour is characteristic of those algæ which grow either in fresh water or in the shallower parts of the sea; the olive-coloured algæ are most abundant between the tide-marks; while the red-coloured species occur chiefly in the deeper and darker parts of the sea.

'As regards perpendicular direction, Forbes remarks, that one great marine zone lies between high and low water-marks, and varies in species according to the kind of coast, but exhibits similar phenomena throughout the northern hemisphere. A second zone begins at low water-mark, and extends to a depth of from 7 to 15 fathoms. This is the region of the larger laminarias and other fuci. Marine vegetation, including the lower forms, extends to about 50 fathoms in the British seas, to 70, 80, or 100, in the Mediterranean and the Ægean sea. Ordinary algæ, however, seem scarcely to exist below 50 fathoms. Diatomaceæ exist in the deep abysses of the ocean and nullipora and corallines increase as other algæ diminish, until they characterise a zone of depth where they form the whole obvious vegetation. . . . . . .

'The vertical range of terrestrial vegetation has also been divided into similar zones in altitude. The relation is called hypsometrical. As we ascend from the plain to the top of a mountain we pass through different belts of plants, to such zones of elevation is of vegetation, the extent and variety of which differ in different countries. When Tournefort ascended Mount Ararat he was struck with the circumstance, that, as he left the low ground at the base of the mountain, he passed through a series of belts, which reminded him of the countries he had passed through in travelling from the south to the north of Europe. At the base the flora was that of the west of Asia; as he ascended higher he reached the flora of the countries on the north of the Mediterranean, then that of northern Europe, and when he reached the summit he found the Lapland plants.

Humbolt found that on all mountains there occurs such a representation of different floras, and that particular alpine forms are found almost over the whole world at a particular elevation. In describing the South American alpine flora he says:—" In the burning plains, scarce raised above the level of the Southern Ocean, we find musaceæ, cycadaceæ, and palmæ, in the greatest luxuriance; after them, shaded by the lofty sides of the valleys in the Andes, arborescent ferns; next in succession, bedewed by cool misty clouds, cinchonas appear. When lofty trees cease, we come to aralias, thibaudias, and myrtle-leaved andromedas; these are succeeded by bejarias abounding in resin, and forming a purple belt around the mountains. In the stony region of the paramos, the more lofty plants and showy flowering herbs disappear, and are succeded by large meadows covered with grasses on which the llama feeds. We now reach the bare trachytic rocks, on which the lowest tribes of plants flourish. Parmelias, lecideas, and leprarias, with their many coloured thalli and fructification, form the flora of this inhospitable zone. Patches of recently fallen snow now begin to cover the last efforts of vegetable life, and then the line of eternal snow begins.'

Madden and Strachy give the following account of the Himalayan vegetation, proceeding from the plains of India through Kemaon to Tibet:—' Ascending, we find forms of temperate climates gradually introduced above 3000 feet, as seen in species of pinus, rosa, rubus, quercus, berberis, primula, &c. At 5000 feet the arboreous vegetation of the plains is altogether superseded by such trees as oaks, rhododendron, andromeda, cypress, and pine. The first ridge crossed ascends to a height of 8700 feet in a distance of not more than 10 or 12 miles from the termination of the plains. The European character of the vegetation is here thoroughly established, and although specific identities are comparatively rare, the representative forms are most abundant. From 7000 to 11,000 feet, the region of the alpine forest, the trees most common are oak, horse-chestnut, elm, maple, pine, yew, hazel growing to a large tree, and many others. At about 11,500 feet the forest ends, picea, webbiana, and betula bhojpatra, being usually the last trees. Shrubs continue in abundance for about 1000 feet more; and about 12,000 feet the vegetation becomes almost entirely herbaceous. On this southern face of the mountains the snow-line is probably at about an elevation of 15,500 feet. The highest dicotyledonous plant noticed was at about 17,500 feet, probably a species of echinospermum. A urtica also is common at these heights. The snow-line here recedes to 18,500 or 19,000 feet. In Tibet itself the vegetation is scanty in the extreme, consisting chiefly of caragana, species of artemisia, astragalus, potentilla, a few gramineæ, &c. The cultivation of barley extends to 14,000 feet. Turnips and radishes on rare occasions are cultivated at nearly 16,000 feet. Vegetation ends at about 17,500 feet, scanty pasturage being found in favoured localities at this elevation; and the highest flowering plants are corydalis, cruciferæ, nepeta, sedum, and a few others.'

'If we examine the vegetation of the mountains of Europe we shall find,' says Balfour, 'a series of similar changes. In the regions of the plains and lower hills of the Alps, extending to 1,700 feet, the vine grows; to this succeeds the zone of chestnuts, which extends to 2,500 feet; the zone of the beech, and of the higher dicotyledonous trees, reaches from 2,500 to 4000 feet; we then come to the sub-alpine region, the zone of coniferæ, extending to about 6000 feet, in which are found the Scotch fir, the spruce, the larch, and the Siberian pine, along with certain sub-alpine forms of herbaceous plants; next comes the alpine region, or the zone of shrubs, extending to 7000 feet, characterised by rhododendron hirsutum and R. ferrugineum, which represent the bejarias of the Andes; finally, we reach the sub-nival region, extending to 8,500 feet, and comprehending the part between the limits of shrubs and the snow-line, where we meet with numerous species of ranunculus, draba, saxifraga, gentiana, primula, and poa, besides other genera belonging to ranunculaceae, cruciferae, caryophyllaceae, leguminosae, compositae, gramineae, lichenes, and musci. On some of the Alps we find flowering plants reaching to the height of between 10,000 and 11,000 feet or more. Schlagintweit found, on the central and southern Alps, at from 10,650 to 11,700 feet, androsace glacialis, A. helvetica, cerastium latifolium, cherleria seloides, chrysanthemum alpinum, gentiana bavarica, ranunculus glacialis, saxifraga bryoides, S. oppositifolia, and silene acaulis. The extreme limit of mosses in the Alps is in general little above that of phanerogamous plants. The last lichens are to be found on the highest summits of the Alps, attached to projecting rocks, without any limitation of height.

'On the Pyrenees the following zones are observed:— 1. The zone of vine and maize cultivation, and of the chestnut woods. 2. A zone extending from the limit of the vine to about 4,200 feet, at which limit the cultivation of rye ceases; here we meet with buxus sempervirens, saxifraga geum, arinus alpinus, ernica montana, &c. 3. From the limit of the cultivation of esculent vegetables at 4,200 feet, to the zone of the spruce fir. 4. From the limit of the spruce fir zone at 6000 to 7,200 feet, characterised by the presence of the Scotch fir. 5. From 7000 to 8,400 feet, is an alpine zone, characterised by the dwarf juniper, draba aizoides, saxifraga bryoides, soldanella alpina, juncus trifidus, &c. 6. A zone above 8,400 feet, exhibits a few alpine species, as ranunculus glacialis, draba nivalis, stellaria cerastoides, androsace alpina, and saxifraga groenlandica.

'There are thus in lofty mountain districts evident belts of vegetation. At the lower part is the region of *lowland cultivation*, where the ordinary cultivated plants of the country thrive. In cold regions this is very limited, while in warm regions it is extended. To this region succeeds that of *trees*. In high northern latitudes, as at 70°, it reaches to between 700 and 800 feet; on Ætna to 6200;

on the Andes to 10,800, and it is marked by escallonia myrtilloides, aralia avicennifolia, and drymis winteri; on the mountains of Mexico to 12,000 feet, and is marked by pinus montezumæ; on the south side of the Himalaya to 11,500, and on the north side to 14,000 feet. On the Pyrenees its limits are marked at about 7000 feet by pinus uncinata, on the Alps at about 6000 feet by pinus picea, on the Caucasian mountains at 6,700 feet, and in Lapland at about 1,500 feet by the birch. Next in order comes the *Shrubby region*, the limits of which in Europe are marked by rhododendrons, which cease on the Alps at 7,400 feet, and on the Pyrenees at 8,332 feet; on the Andes it is limited by bejaris and shrubby compositæ, at a height of 13,420 feet; on the south side of the Himalaya, by species of juniper, willow, and ribes, at an elevation of 11,500 feet. In Lapland, species of willow and vaccinium, with the dwarf birch, reach 3,300 feet. The next region is that of *grasses*, which on the Andes and the Himalaya extends to between 14,000 and 15,000. Finally, we come to the region of *Cryptogamic plants*, which extend to the snow-line, lichens being the last plants met with.

' In contrasting the zones of altitude with those of latitude, Meyen gives the following regions of alpine vegetation:— The region of palms and bananas (equatorial) extending from the sea level to 1,900 feet; the region of tree-ferns and of figs (tropical) 1,900 to 3,800 feet; the region of myrtles and laurels (sub-tropical) 3,800 to 5,700 feet; region of evergreen dicotyledonous trees (warm temperate) 5,700 to 7,600 feet; region of deciduous dicotyledonous trees (cold temperate) 7,600 to 9,500 feet; region of abietineae (sub-arctic) 9,500 to 11,400 feet; region of rhododendrons (arctic) 11,400 to 13,300 feet; region of alpine plants (polar) 13,300 to 15,200.

On the Table Mountain range there grows luxuriantly the silver tree (*Leucadendron argenteum*), the sparkling in the sunshine of the silver-like leaves of which charms the eye of visitors. It grows at a certain altitude, not in a continuous zone, but in extensive patches, some higher, some lower, all within a definite range, beneath which, only a little way, it is found impracticable to cultivate it.

Such facts as have been cited seem, together, to warrant the supposition that atmospheric and oceanic pressure may possibly have some influence in permitting and preventing the growth of certain plants within and beyond a certain range. It may be that we can attribute to this but little influence in comparison with the modification of light, and still more the modification of heat, produced by altitude above the level of the ocean, and depth below its surface. But be the operation of the influence what it may; and be it the case that it is only in the case of such modification of pressure, light, and heat, as are strongly marked, and are incidental to difference in altitude, that it arrests attention, such differences as are indicated in most localities by differences in barometric phenomena have come to be valued in investigations of the natural history of these plants.

The following is a summary of observations of the kind made in Norway, for which I am indebted to Dr Broch's report:—

In estimating the barometric pressure of the atmosphere the calculation is made of what it would be in the locality at the level of the sea. The mean annual pressure attains its maximum towards the south-west of the Scadinavian peninsula. It has its minimum in the region from the Lofoden Islands to the North Cape, descending from 758 mm. at Mandal to 753 at the place last mentioned. At Christiania the mean pressure of the air is 757.7 mm. Twice in the course of the last forty years it has been 786 mm., and it has been seen as low as 721 mm.

In winter the distribution of the pressure accords as a general rule with what it is on the average for the year. Greatest towards the south-east, the pressure decreases towards the south-west, the isobaric line passing through localities with equal barometric indications of atmospheric pressure, enclosing at Iceland a zone wherein it is the lowest of all; but the Scandinavian peninsula occasions a little disturbance of the regularity of its distribution.

In summer the distribution of pressure is almost the contrary of what it is in winter. Along the axis of the peninsula, from the central mountain mass of the Dovrefjeld to the North Cape, there exists a region in which it is comparatively low, with an isobaric line of 655 mm., and parallel to this are other isobaric lines following the direction of the coast, so that, in the central portion of the peninsula, there is in summer a minimum of pressure begirt by a greater pressure along the coasts.

The isobaric lines for the winter and for the year are remarkably accordant with those of the anomalous thermometric distribution, great excess of heat corresponding with a feeble pressure of air, and *vice versa*. In summer the same thing is observable, but it is less striking.

In the annual variations there are some small inequalities which do not disappear under protracted observation. But saving these exceptions, over the whole of Norway the greatest pressure is in the month of May. The lowest pressure is seen on the west coast during the months of winter. At the Skagerack it is seen in spring and summer, in the eastern part of the country in July. This variation is least in the east, where it is 3·5 or 4 mm.; it increases towards the west, till at Hammerfest it has attained to 11 mm. This variation is mainly a consequence of the lower pressure in winter.

The diurnal variation during winter is pretty uniform in the eastern and western parts, with a maximum of 4 mm., the maximum occurring at Christiania before mid-day, at Bergen at noon; the minimum occurs between 2 and 4 A.M.

During summer, on the contrary, the diurnal variation is proportionally very great in the eastern part of the country, where it amounts to 1·22 mm., while on the west coast it is very little, amounting to 0·27 mm. only. The maximum occurs at Christiania at 7 A.M., at Bergen at noon, in both places the minimum is at 6 P.M. At Christiania the lowest amplitude is 0·5 mm.; the maximum occurs a little before mid-day, the minimum at 6 P.M. The secondary noturnal maximum and minimum there are very marked.

The Highlands of Norway are more frequently high-lying plateaux designated fjelds,* than in outline what is generally suggested to an English reader by the designation mountains; but mountains they are.

'In the northern district of Scandinavia,' says Forbes, 'where the theory of a ridge is in some respects less inaccurate than in the south, its insufficiency was clearly discovered by the difficulty or impossibility of defining the line of demarcation between Norway and Sweden by that of a continuous watershed. Such a ridge, if it exist at all, must be held in some cases to run up to the very coast of Norway, or even beyond it, into the islands; in other places it dies out altogether, and is resumed with a change of direction.

'The present boundary between Norway and Sweden was defined by a joint commission of engineers in the middle of the last century, and is represented on nearly every map as the exact direction of a slightly zig-zag chain of mountains called the Kjölen, or Kœlen. This is assumed in most maps to be prolonged along the border of the two countries considerably to the south-east of Trondhjem (Drontheim), and it was even long maintained that a mountain mass existed there of prodigious elevation, from which a great many rivers, particularly the Glommen, the Gota, and the Dal, take their rise. The height of this fabulous mountain was even assumed to be 12,000 feet. It is, however, only a slight and lower extension of the Dovre fjeld beyond the deep valley of the Glommen, and its greatest height does not amount to 5000 feet.'

---

* By Dr Broch there is given an orographic table, in which is supplied tabulated information in regard to the different mountain chains and mountain plateaux of Norway, the principal and the secondary regions, and divisions of these, and the names and altitudes of between five and six hundred of the more elevated peaks.

In a paragraph preceding this he says:—' Thus the general surface of the country is in reality composed of elevated and barren table-lands. The proportion of arable land (land which might be tilled) to the entire extent of Norway, is not, according to the competent authority of Professor Munch, more than 1 to 10; and if we exclude a few local enlargements of the plains near the capital, it would not even exceed 1 to 100. By a rude estimate in Professor Keilan's map, I find that the portion of the surface of Norway south of the Trondhjem fiord, which *exceeds* 3000 feet above the level of the sea, amounts to very nearly 40 per cent. of the whole; and when it is recollected that only one summit exceeds 8000 feet, and that the spaces exceeding 6000 are almost inappreciable on the map, it will be more clearly understood how completely the mountains have the character of table-lands, whose average height probably rather falls short of than exceeds 4000 feet.'*

Detailed information in regard to all the mountain systems is supplied by Dr Broch. Passing over his statements in regard to those in western and northern Norway we find him stating in regard to the mountain chains, plateaux, and masses, in southern Norway, after having described them with similar details, that of these, the Drontheim plateau, the Dovre fjeld, &c., constitute a remarkable whole in regard to physical and orographic character. Cut up almost to infinity, belonging at once to the east, to the west, and to the north of the country, they occupy almost half of the portion of Norway situated between the country to the south of Drontheim and of the valley of the Levanger to Jemtland in Sweden. These mountains thus cover very nearly the fourth part of the whole superficies of Norway. They reunite in a continuous mountain mass, the whole of which by the south of this limit rises more than 600 metres, 2000 feet, above the level of the sea.

---

* These estimates refer to Rhenish or German feet, which are about 3 per cent. larger than English feet.

All of this, which is situated between 600 metres and 1000 metres, say 3,500 feet, is still to a great extent, especially on the east side and in the lower half, covered with coniferous trees, *Pinus abies* and *P. sylvestris*. In the higher portion of this zone the conifers give place to the birch, *Betula glutinosa*, which, especially in the slopes facing the south, and in places sheltered against the north winds, attain an elevation considerably above 1000 metres, say 3,500 feet, above the sea level. The juniper, *Juniperus communis*, rises higher on dry land; and at a still greater elevation are found in moist places smaller species of willows, *Salices*, and the dwarf birch, *Betula nana*. They attain in sheltered places an altitude of 1500 metres, 5000 feet. Higher still vegetation is represented by different grasses, and carices, and sedges, beyond that by mosses, and finally by lichens which grow upon the bare rock.

It may further be remarked that the level surface of the mountain plateau, where they are not wooded—and, as has been stated, the forests prevail over a portion of the zone situated between 600 and 1000 metres in altitude—consist in part of naked rock covered only with mosses and heath, in part of marshes to which marsh berries in some places give it a colouring, in part of bogs which yield to the foot, and in part of herbaceous turf and verdant flower-decked slopes. On other parts we meet with ancient moraines composed now of a soft and deceptive clay covered with schistose fragments, in which the incautious traveller may sink to the knees. In another part are projections of rock, bare, or covered with great heaps of stones, or, it may be, even to the end of summer, with large and small sheets of snow. Névés and glaciers appear in many places, and cover sometimes a continuous bed of some hundred square kilometres, say of hundreds of square miles. Thus is it in the region comprised between the Sogne fiord and the Nord fiord, in the prefecture of Nordre-Bergenhus, and in that of the Folgefonn to the south-east of the Hardanger fiord, in the prefecture of Soendore-Bergenhus. The highest part, the most wild, the most bare, the most snowy, the most wintery, the most desert, the most gloomy of this plateau, stretches along the western declivities of the precipice.

From these mountain chains and elevated plateaux in Southern Norway come down systems of mountains of lesser attitude, dominated in certain places by elevated summits, but which, as a general rule, form wooded ridges of little elevation. And last of all come the low lying lands, but which only in certain places stretch themselves out in great plains.

In all the Vesteraolen, Lofoden, and Sonjen groups of islands, lying considerably to the north of Drontheim, the mountains—lofty, sharp-edged, and tooth-shaped, or serrated, often conical or pyramidal—have a precipitous descent towards the sea, leaving but rarely a narrow littoral strip of turfy soil. On some points only do the mountains recede far enough to give rise to small valleys, generally filled with marshes. The interior of the large islands is in like manner filled with bogs, which are often, as is the Dvergbergmyr, in the Andoe, of considerable extent, and covered with cloudberry (*Rubus chamaemorus*). The summits of the mountains rise often to from 800 to 1,300 metres, or 3,700 to 4,350 feet. On the island of Andoe, the most northern of the Vesteraolen, the mountains are lower and the summits more rounded.

On many parts of the Lofoden Islands and Vesteraolen Islands on the coast looking towards the Arctic Ocean are found remarkable bird mountains called *Nyker*, inhabited by marcareax (*Mormon articus*)—common penguins (*Alca Torda*). Large guillemots (*Uria troile*), and three-toed mews (*Larus tridactyllus*). The *Nyker* are composed of steep pyramidal mountains, which shoot up directly from the sea, without any cincture of rocks, so it is only after a long calm, and when the wind blows off the land, that it is possible to land upon them. They are often covered with a layer of sandy vegetable earth of the thickness of from 50 to 60 centimetres, 20 to 24 inches, the surface of which presents the appearance of clods disposed in irregular steps, covered with a vigorous herbage. Each of these clods is mined by passages where the birds make their nests. At some little distance from the sea-shore these passages are continued, sometimes straight, sometimes sinuously, but in a direction very nearly horizontal, to a depth of from 1 to 2 metres, 40 to 80 inches; they are about 30 centimetres, 12 inches, in diameter.

The *Nyker*, the inhabitants of which are to be reckoned by millions, are at the moment when the birds quit their nests so surrounded by countless swarms, that at a distance they appear as if enveloped in clouds or in a crape-veil. There is heard afar off a humming sound, as from a swarm of bees, and when the midst of the birds is reached, the noise is altered to a roar, like that of a violent storm or tempest. The *Nyker* appear there riddled by white spots in perpetual motion, or as if seen through a dense fall of snow, occasioned by the movements of the birds coming and going. The most remarkable of these *Nyker* are found in some heights which shoot up directly from the sea in the neighbourhood of the isle of Rost, near Malnæs, on the west coast of the island of Lango, and on the west coast of the island of Ando. The *Nyker* are inhabited by birds of passage, which quit them in the months of August and September, to go further into the region of the Arctic Ocean, whence they return in March and April. To take them in the subterranean mines where they are found in their nests, there are employed dogs like the turnspit, trained for the purpose. Part of the birds are eaten salted, and money is made by the sale of the eggs. These birds there take the place of vegetation.

The mountain land stretching between the Porsanger fiord on the north-west, and the Varangar fiord on the east; and on the south-west the littoral mountain chain of West Finmark is the Finmark plateau, which, in contradistinction to the lands already described, may be described as a flat country. Its mean elevation is about

310 metres, 1040 feet, above the level of the sea. It extends into Finnish Lapland on the south-east, as far as the great Finnish lake Enare, which measures 2530 square kilometres, and is 150 metres, or 500 feet, above the level of the sea.

This vast plateau is from Vardo to its south-east boundary, between Kautokeino and Finland, 350 kilometres long; and the Norwegian portion covers an area of nearly 40,000 square kilometres. It is traversed from south to north by two considerable rivers. The western of these is the Attenelv.

The plateau of Finmark is traversed from south to north by two considerable rivers. The Attenelv on the west is 160,530 kilometres long. It flows into the Attenfjord. The Tanaelv—that on the east—is formed by the junction of the Karasjoka and the Anarjoka. Calculated with the length of the latter, the Tanaelv is 275 kilometres long, and falls into the sea on the Tana fiord. The Attenelv and the Tana also take their rise on the Russian and Finnish frontier, and flowing in valleys of gentle declivity, almost join one another by the valley of the Jetsjoka, one of the affluents of the Anarjoka. The church of Kautokeino, situated only 40 kilometres from the source of the Anarjoka, is 290 metres, 880 feet, above the level of the sea.

In the flat valleys which form the mountain plateau of Finmark we met with some stunted birch trees; in the valleys somewhat deeper we meet also with pines; but otherwise the plateau is bare, or covered only with reindeer moss (*Cladoma rangi ferina*), and strewed with rounded stones and large fragments of rock. These last are often through stretches, dozens of kilometres in length, the only objects on which the eye can rest. They are also well known, and serve as *reperis* for travellers who cross the plateau. This is done exclusively by means of reindeer. The plateau is, so to speak, sown with small lakes. The heights which rise above the level of the plateau reach 600, and exceptionally, 1000 metres of altitude, 3,333 feet.

Of the 316,580 square kilometres which constitute the total area of Norway, 37,000 kilometres have an altitude of more than 1000 metres, 3,333 feet; and about 91,000 from 800 to 1000 metres, 1,600 to 3,200 feet, above the sea. The mean altitude of the whole territory may be estimated at 490 metres, 1,650 feet, above the level of the sea.

From this it may be gathered that while their are mountains—isolated mountains, and even chains of mountains—rising above the level of high-lying lands, these plateaux impart to the superficial contour of Norway its characteristic outline; and in accordance with this we hear more—a great deal more—of fjelds than of hills.

There are not awanting mountain crests of pointed rocks, and rounded nut-like mountains, and mountains and islands which may be described as grotesque in their shape. One island, Torghatten, is perforated by a natural tunnel, which from the sea appears like a bright loop-hole in a dome of rock; an island near the Arctic Circle has been named the Hestman or Horseman, from its resemblance to a mounted knight; and other forms are as well defined, though more difficult of description by reference to well-known objects. But the general character of the mountain region is an extensive level plateau of high altitude cut up by ravines, the sides of which are almost perpendicular.

Of the fjeld, as it is called, I find it difficult to convey a definite and accurate conception. As a preparation I may speak of such at the outset as upland moors; but this would not prove exactly descriptive of the Dovrefjeld, the Hardangerfjeld, and others of which the traveller hears so soon as he enters the country, and which he is likely to visit if he has come to see its wonders and enjoy its scenery; and yet I feel disposed to retain it as accordant with the idea I wish to convey as a preparation for a more precise description.

K

The designation fjeld is, according to Forbes, given to extensive *plateaux*, or table-lands of great breadth, and generally more or less connected together, though occasionally separated by *deep* but always narrow valleys. As seen from the lowlands they appear mountain ridges—mountain ranges they are, but they can scarcely be called with propriety mountain ridges. 'They are often,' writes Forbes in his volume entitled *Norway and its Glaciers*, 'interminable wildernesses, undulating, or varied only by craggy heights devoid of majesty, rarely attaining the snow line, but spotted over with ungainly patches of white.'

Of the Dovrefjeld graphic descriptions have been given by Forbes and also by Bayard Taylor, as well as by many other tourists.

'They are often so level that upon what may almost be called their summits a coach-and-four,' says he, 'might be driven along or across them for many miles did roads exist, across which the eye wanders for immense distances, overlooking entirely the valleys, which are concealed by their narrowness, and interrupted only by undulations of ground, or by small mountains which rise here and there with comparatively little picturesque effort above the general level.'

And in another connection he mentions that the forms of the Norwegian mountains, contrasted with the Alps, have been aptly enough compared by Wittich—the former to the embrasure of a parapet, the latter to a ridge and furrow roof, the depressions in the former representing the profound gorges which intersect the rocky plateaux; in the latter the usual alternations of mountain and valley.

Of the ravines which cut up the plateaux a picture has been given in connection with what has been told in regard to the Marie Stegen and the Rinkan Foss [ante p. 24]. The perpendicular precipice so frequently characteristic of the ravine is not unfrequently met with in other scenes. In illustration I may cite the following account of the Horningdalsrokken, a peak crowning one of the finest precipices in Norway. Mr John P. Campbell, the writer of a little book entitled *How to See Norway*, states that he was the first Englishman who reached the top, and he gives the following account of his ascent:—'I arrived 'at Haugen on the evening of the 27th July 1866. Lars Elias, the station-master, gave me some porridge and a bed, and next day we two started about 5.30 A.M. in a cart. Our drive was some three and a-half miles up the valley to a *saeter*, where we left the cart-horse and cart, and the rest of the way was on foot. Two miles or so brought us near to the head of the glen, eventually getting clear of the forest, and to a green knoll which overlooked a tarn. This water was probably 1000 feet above the level of the sea; and almost vertical from its margin rose the peak we had in view—a straight wall of rock between 3000 and 4000 feet high. The summit, seen from below, appeared to terminate in a ruined tower; but it was not so (as I afterwards found), being in reality a ridge, of which we only saw the end.

'The ascent from where we stood looked uninviting enough; but Lars had been up several times before, and never hesitated about the route. We followed a corry sheltered by this wall of crag, up to a col, or slack, which it took us one and a-half hours to reach. It was very stiff climbing; and from the steepness and slippery nature of the ground, the descent of this portion on our return was quite as slow. For a long way up there was verdure, including ferns and bilberries, which decked the slopes leading between fjeldhammer—as crags forming terraces across a mountain side are called—but as we approached the col this disappeared. We were now on the upper part of a field of névé, from which flowed a glacier down the reverse side of the fell. Gently rising now in a direction parallel to the glen, we traversed the névé, the ridge being above us on our left. The snow was just right for walking on, and there was no difficulty in winding round to its junction with the rock at the farther and more accessible end of the ridge. The edge was very narrow, so much so that on one part I adopted the crawling system, like a bear. It sloped up gently to the top, and then continued nearly horizontal for some way. The whole ridge was bare of snow, forming a crest on the mountain like the comb of a cock. We were obliged to follow the edge of it owing to the smoothness of the craggy slope on the left. As to the other side, one might have measured it with a plummet. According to a legend, a very long time ago a trold, or giant—who resided on the top--used to sit there and fish the tarn below by throwing down a line. A cairn marked the highest point. The view was wonderfully wild. Following the way we had come, we reached Haugen at 3.30 P.M.'

CHAPTER XIV.

TEMPERATURE AND ALTITUDES OF SNOW-FIELDS AND GLACIERS.

At altitudes varying in different localities according to their latitudes, we reach a zone of perpetual—or, rather, as it should be called, perennial—snow. The snow-fields beyond are known as névés, while the fringe of ice begirding their lower edge are known as glaciers. In Norway we find both in the higher-lying plateaux, the remains, it may be, of a far more extensive Arctic snow-field than that which now exists in polar regions, one which, with snow and ice, during what in geology is called the glacial period, covered extensively the whole of Europe.

While it is the case that difference in soil and difference in atmospheric pressure at different altitudes in a mountainous country are not to be altogether ignored in considering causes of the geographical distribution of plants, it is chiefly as an indication of temperature at which different kinds of trees are found at different altitudes that it is deemed of importance by the student of forest science. In this connection the line of perpetual snow is supposed to supply him with a valuable indication of a temperature which never falls much below the freezing point, from which he learn much by ascertaining how near to this may different species of trees can grow and flourish.

In Norway in the zones of the oak, and of the birch, and of the cultivated fields, the temperature is moderate. In the midland districts the cold is more severe; but there pine and fir forests of boundless extent rise on high stony ranges, intersected with plains and valleys of meadow and cultivated land, and dells where the willow and the alder vegetate in great luxuriance. 'And here,' writes one who spent ten years in Scandinavia, and has chosen the pseudonym of *An Old Bushman,* author of *Bush Wanderings in Australia,* ' are vast morasses, many of which can never be traversed by the human foot, rivers and inland lakes of every size, fringed with the reed, the bulrush, and the candock, and thousands of acres of low meadow land, covered with thick, coarse grass. It is here that the British naturalist begins to meet with rare and new specimens, and it is here that the eye of the traveller first gazes on the fine scenery of the north: and more beautiful scenery than Scandinavia displays during the summer months it would be hard to find. I have wandered over many lands, but scarcely ever saw a European landscape to vie with this.'

In the very far north the appearance of the whole country becomes gradually more wild and rugged, and high mountains and barren fells, covered with perennial snows, rise above the limits of vegetation, and tower over the forests which skirt their base.

The great névés, or snow-fields of Norway, differ from those of the Tyrol and of Switzerland, in that they do not lean upon collossal mountain masses, and naked summits which denominate them. They occupy, on the contrary, the elevated parts of the great rocky plateaux, on the slopes of which they spread out their toothed branches; and in this respect they resemble still the névés which cover the interior of Greenland.

Of the snow-line, or that above which the snow lies constantly, Forbes writes:—' The occurrence of perpetual snow at a certain height above the sea, in even the warmest regions of the globe, has in all ages excited the curiosity of geographers and naturalists. Regarded at first as a very simple indication of the depression of temperature as we ascend in the atmosphere, it has been carefully studied and applied (often erroneously) to the determination of climate. Closer examination has shown that the presence of perennial snow, in other words, a predominance of all the causes tending to its accumulation over those which tend to its waste or fusion, is indeed a very complicated fact, and cannot be taken as the simple expression of any one of the elements of the climate. The snow line is far from having invariably a mean temperature of 32°, as was at one time supposed. Under the equator it is about 35°; in the Alps and Pyrenees about 25°; and in latitude 68° in Norway it is (according to Van Buch), only 21°. Yet there are regions, both in the extremity of Siberia and in Arctic America, of which the mean temperature is below zero of Fahrenheit (as, *e.g.*, Melville Island). And it is quite established, on the concurrent authority of those best acquainted with those regions, *that nowhere in the northern hemisphere does the snow-line attain the low level of the sea.* The explanation is to be sought principally in the intensity of the summer heat during the period of perpetual day, which effectually thaws the soil, though only to a trifling depth, and raises upon its surface a certain amount of brief vegetation, suitable for the support of Arctic animals.

'Another cause affecting exceedingly the level of the snow-line is the amount of snow which falls. The interior of continents being far drier than the coasts, the snow to be melted is a comparatively slight covering. The snow-line on the *north* side of the Himalaya is at least 3000 feet higher than towards the burning plains of Hindostan. This is chiefly due to the excessive dryness of the climate of Thibet. In like manner five times less rain falls on the coast of the Baltic than at Bergen. All this confirms the excellent generalisation of Von Buch that *it is the temperature of the summer months which determines the plain of perpetual snow.* It is thus easy to understand why the mean temperature of the snow-line diminishes towards the pole, because for a given mean temperature of the whole year, the summer is far hotter in proportion. Also, places at which the temperature of the summer is low are those which have a moderated or coast climate; but there also the fall of rain and snow is most abundant, whilst in excessive or continental climates the precipitations are comparatively small. Thus, to take one illustration, in Iceland snow lies all the year at a height of only 3,100 feet, whilst in Norway, on the same parallel, the snow-line would approach 4000 feet.

'The same general principle holds good in the Southern Hemisphere; its temperature on the whole being greatly inferior to that of the north (though the extremes are less). It acts towards the rest of the globe in some measure as the refrigeratory of a great distilling apparatus (as some one has correctly observed), and its higher latitudes are the seat of almost continual storms and fog, of which the climate of Cape Horn is a familiar example. Summer can there hardly be said to exist, and the snow-line is proportionally low. According to Sir James Ross, the first authority of his time on this subject, the snow-line does reach the level of the sea in the Antarctic regions at a latitude between 67° and 71°, under which forests still grow in Norway, and even corn in some sheltered places.'

Forbes proceeds to give numerous estimates of the altitude of the snow-line in different parts of Norway, which may be of great value in the study of meteorology, but what we require is simply the substance of the whole; and a generalisation of the observations made or followed by him seems to show:—

First. In latitude 60° to 62°, the snow-line at a short distance from the coast may be considered to be 4,300 English feet, or thereabout; *secondly,* in the same latitude, 60° to 62°, towards the centre of the country, it rises to 5,300; *thirdly,* in latitude 67°, in the interior, it is only 3,500 feet; it is not much lower on *insulated* summits in latitude 70°, but on the coast it is as low as 2,900 feet.

It is observed that the summer isothermal line shows a marked tendency to run parallel to the peninsula, and to this this trifling effect of latitude is in part attributable.

Von Buch has remarked that in Norway and Lapland the planes of vegetation of the pine and birch run nearly parallel to the plane of perpetual snow,—the intervals, as observed by him at Alten, being given by the following table of limiting heights of vegetation above the sea:—

VEGETATION IN LATITUDE 70°.

The Pine (*Pinus sylvestris*) ceases at	237 metres	=	780	Eng. feet.
The Birch (*Betula alba*) ,,	482 ,,	=	1580	,,
The Bilberry (*Vaccinium Myrtillus*), ,,	620 ,,	=	2030	,,
The Mountain Willow (*Salix mirsinites*), ,,	656 ,,	=	2150	,,
The Dwarf Birch (*Betula nana*), ,,	836 ,,	=	2740	,,
The Snow-line, ,,	1060 ,,	=	3480	,,

From the growth of the birch he has estimated the level of the snow-line in the islands of Qualoe and Mageroe, though neither of these rise to the requisite limit. It is probable, however, that the direct sea-blast to which those bare rocks are exposed, must act chemically upon vegetation in a way which would render the deduction of the snow-line considerably doubtful—which doubt is confirmed by the more recent determination of the snow-line on the island of Seiland, opposite to Qualoe.

Von Buch estimates the interval between the limit of the birch and perpetual snow at about 1,870 English feet throughout Norway; Wahlenberg estimates it at 1,960 English feet; which probably represents best the results in higher latitudes.

And as a guide to fill up the gaps of direct observation, he adds, some determination of the limit level of the common birch in Norway, taken chiefly from the *Gœa Norvegica*, from *Naumann's Travels*, and from the observations of Wahlenberg and of Smith, the Norwegian botanist. These he has given in tabular form, representing the estimated altitude of the snow-line in twelve different localities, ranging from 59½° to 71° 2', adding to the mean limit of the birch 900 feet, as the limit of the birch and perpetual snow, and in six of the cases the altitude of the observed snow-line, which shows a general accordance, and, in some cases, an approximate conformity, of the one with the other.

CHAPTER XV.

MECHANICAL ACTION OF GLACIERS.

BESIDES the interest which attaches to névés and glaciers of Norway as indications of temperature at high altitudes, they possess an interest for students of physical geography as remains of yet more extensive névés and glaciers, which in bye-gone ages exerted a powerful influence in cutting up and carving, or rather graving, the contour of the country, creating the wild wonders of its features which make it so attractive to tourists who are in quest of the wild, the magnificent, and the grand in nature.

As the forests of Norway are the remains of forests which once covered more or less entirely the whole of Europe, so are these névés and glaciers the remains of a sheet of snow and ice which once covered extensively Northern Europe, if not the whole continent and lands beyond it; and the existing fiords may be looked upon as having been to a great extent, if not entirely, the creation of that far-reaching sheet, as under the superincumbent pressure of the mass, it sought, here and there, to find a way to the lower level of the ocean bed, ploughing, and undermining, and sweeping along with it all *débris* as it advanced on its resistless course.

The interest which may be awakened in looking upon the wild outlines and contour of these fiords in proceeding from the coast to the glaciers on the elevated plateaux of the interior, may be intensified, if they be contemplated under a dominating influence of such considerations.

With this view I would bring forward what has been told of one of these fiords by the experienced traveller whom we took as our guide from Christiania to the

Marie Stegen, who, we found had disappeared while we were musing on the exciting scene, and on the romantic incident to which is attributed the name which it bears, but of whom I said we might possibly overtake him at the Sogne fiord.

The Christiania fiord, of which mention has been made in the opening chapter, which treats of the general features of the country, is a continuation of the Skagerrack, and resembles closely the rocky shore scenery through which steamers make their way from Christiansand to the capital. A very different appearance is presented by the Sogne fiord, and this is more characteristic of the fiords of Norway.

The Sogne fiord may be visited most conveniently by steamer from Bergen, for which port steamers sail frequently, if not every day, from Christiania.

M. Du Chaillu, writing of his visit to this fiord, tells:—

'Of all the fiords of Norway none can rival in size, grandeur, bold outlines, weird and sombre landscape, the magnificent Sogne. No tourist should fail to sail upon its waters. The entrance, which is formed on the west side, by the Sulen Islands and others, is at about 61°, and the main course winds its way inland almost directly east. . . . The average breadth of the Sogne varies from three or four to about two miles, and its length, in a direct line, is over three degrees of longitude, or a distance of about eighty-four miles with its windings.

'There are several lateral branches, extending north and south, besides deep bays or coves. On the northern shore are the Vadeim and Fjaerland, the latter fourteen miles long, the Sogndal ten miles, and the Lyster twenty-four miles; on the southern shore are the Brekke, the Arne, and the Aurland, the latter being sixteen miles long, with its branch, the Naeroe, about six miles. No description can give to the reader an adequate idea of the magnificence of the scenery of these narrow lateral fiords of the Sogne. . . . The route to the Sogne fiord is among so many islands that it often seems as if you were sailing on a river. The scenery at times is extremely fine. The greater part of the country is uninhabited; now and then the sea is so completely land-locked that it appears as if the journey was ended, when suddenly comes into view an opening, and another broad expanse of water stretches in the distance; the channel is sometimes so narrow and tortuous that the vessel almost touches the rocks. . . .

'In about six hours from Bergen the entrance to the Sogne is reached, where it is six or seven miles wide. Skirting the southern shore, you pass a grand mass of rocks. The Sognefest—the Castle of the Sogne—is very bold in its outlines, and apparently forming two sides of a square. The scenery spread before the traveller is superb, a panorama ever changing in its views of snow-topped mountains: in the north are the Justedal glaciers, towering mountains in the east, in the south the snow-fields of Fresvik. The vegetation improves as you penetrate inland; the bases of the mountains and hills are clad with woods. . . . . The valleys by the fiords are often quite fertile and well-cultivated, contrasting singularly with the barren mountains which surround them. From the water they appear to form an oval basin with a ravine at the end, towards which the mountain sides slope gently, evidently hollowed by the agency of ice and water. Sometimes two ravines enter the valley, like radiating branches. At the base of the mountains the terraces rise one above another to the number of three or four.

'At about sixty miles from its entrance the Sogne seems suddenly to end at the base of a high mountain; it sharply turns northward, and the island of Kvamsoe is passed, and a few miles further the main fiord runs once more eastward, while to the north is the entrance of Fjaerland, the first large branch of the Sogne.

'The steamer stops at the thrifty hamlet of Balholmen, opposite to which is Vangsnaes, the scene of Frithiof's Saga. Sombre is the Fjaerland, with its mountains, glaciers, and its wild scenery. Streams, fed by the melted snow and the ice, run down on every side. In the mountains above are the Langedal and the Bjorne glaciers, rising to 4,500 and 4,780 feet above the sea; a little farther north, on the west side, are the Svaere and the Vetle fiords, between mountains, the highest of which, the Oatneskri, rises 5000 feet—nearly a mile, which is 1,760 yards, or 5,280 feet. At the end of the Vetle fiord there is a road of a few miles, leading to the great ice-field of Justedalfonn. As you sail farther inland, still higher mountains loom up on both sides of the fiord—the Melsnipa, 5,620 feet; the Gunvords and Stendale glaciers, 5,200 feet. The water is of a peculiar opaque light green, showing the effect of the numerous streams from the ice. Three valleys diverge from the lowlands at the end of this fiord. The first, the Suphelle, is a long narrow ravine, enclosed between rugged mountains; its glacier, about four miles from the sea, is fed from the side of another with which it has no direct communication, the masses of ice falling from a height of between two or three thousand feet. . . . In the year 1868 a large number of avalanches occurred in different parts of the country, occasioning loss of life and property. On the Fjaerland, on the west side, one descended of such a size that it formed a bridge over the fiord—at that point 5000 feet [nearly a mile] wide—upon which the people crossed. If I had not been told this by several trustworthy persons, I would not have believed it, so incredible does it appear.

'Leaving the Fjaerland, and again ascending the Sogne fiord, the scenery becomes more cheerful—woods, fields, meadows, and hamlets, are far more numerous: at the base of the mountains the woods crowning even some of the lower hills. Here is the hamlet of Fejos, while the Fresvik snow-field, rising 5000 feet, towers over all. . . . Two streams from the Grindsdal and Henjumdal—two valleys a few miles apart, both formed by the Gunford glacier, 5000 feet above the sea—empty into the sea here, and give water-power to numerous grist-mills.

'A few miles farther up on the northern shore is the

Sogndal fiord, with its weird scenery, its fruitful tracts, and transverse valleys, over which farms are scattered. . .

'From the Sogndal the scenery of the Sogne is superb. On the northern shore rises Storehog, 3,830 feet; opposite, Blejen, 5,400 feet; and the fiord between them is about two miles wide, and 2,900 feet deep. Many of the mountains rising from the fiord are torn; in some places birch, fir, or pines, are seen to a great height; and a solitary farm, a saw or grist-mill meets the eyes. Fifteen miles above the Sogndal fiord, on the northern shore are the small hamlets of Lower and Upper Amble, and Kaupanger church. These are situated on the shores of a lovely bay of oval shape. The lower hills slope gently towards the sea, and are clad with woods to their very tops; while groves of different trees, the elm, the linden, the birch, and other trees, grow here and there. Two beautiful streams fall into the sea, and on their banks are little grist-mills. Meadows, yellow fields, and patches of potatoes were scattered round the farms. On a sunny day the place is exquisitely beautiful. How many of these picturesque spots one finds upon the fiords; they burst upon you when least expected. A little farther, entering the Lyster fiord, one beholds a beautiful and extensive panorama of mountains and water. Snow and glaciers meet the eyes in the higher regions; while a farm, a hamlet, or a church, shows that men live by the sea in the midst of this grand and stupendous nature.

'Some ten or twelve miles inland, on a promontory on the eastern shore, is Urnaes, from which an excellent view of the fiord presents itself, with its ranges of hills and spurs coming down to the sea. On the western shore opposite Urnaes, is Solvorn, picturesquely situated in the hollow of the mountains.

'We are here amongst mountains and glaciers, and waterfalls are not awanting. At the mouth of the Lyster we enter Aardal, a continuation of the Sogne, and its most eastern extremity. At its entrance rises the Bodlenakken 2,990 feet, and on the opposite side the Boermolnasse, 3,860 feet, with still higher mountains beyond them.

'In the wild valley, which is a continuation of the fiord, at a short distance inland,' he goes on to say, 'is a picturesque lake, whose waters are of a deep green colour. . . The Stigebjerg mountain rises perpendicularly from the lake, with a wild waterfall plunging in white foam from a towering height. . . . Towards the middle of the lake the scenery is superb, and looks wild and weird. In one part the gigantic mass of rocks falls abruptly into the water, and a little further on a grand fall—Hellegaard-Foss—tumbles in white foam from the heights above, and looks whiter on account of the sombre nature of the rocks. Perched high up are several saeters, one of which is called Kvenli. Soon after came in view from behind another white mass of foaming water the Stige-Foss, which had been hidden from our view.

'Looking backward towards the fiord a wild spectacle greets the eye, and one cannot realise or believe it is the same country just passed; towering mountains and wild ravines are seen in every direction, and the yellow leaves of the birch and grass look beautiful. Near the upper end of its northern shore is the Nondal valley, with farms pitched 2000 feet above the water. At the head of the lake the valley of the Aardal takes the name of Utland, which leads to the Vetti-Foss—the waterfall for which we are now bound. It runs almost parallel with the Lyster fiord, separated from it by masses of mountains, about twenty-five miles wide, culminating in the Horunger 7,620 feet high, and surrounded by glaciers. On the eastern side the mountains rise to a height of 6,500 feet, and its lakes and torrents afford the artist and the lover of mountain scenery unfailing and ever-changing sources of delight.

'There is a neat farm called Noen, where one can find comfortable quarters. At a short distance from the house a spur of the mountain covered with fir seems almost to bar the way; but beyond this is a beautiful dale with a few farms, looking like an emerald gem. This lovely spot is about an English mile in length. From thence the valley narrows itself into almost a ravine, strewn with fragments rended from the mountain sides, and lined with occasional terraces. Passing the farm of Svalheim, you reach the Hjaelledal-Foss, a superb cascade falling in a sheet of foam from a height of seven or eight hundred feet, and then the Hagadal-Foss, nearly as high. The river below is spanned by a frail, narrow bridge, composed of two or three fir logs; and on the other side there are a few fields of barley, and a patch of potatoes. . . . Afterwards the Utland becomes very narrow, and almost obstructed by huge rocks—masses of rock which fall every year from the mountain, against which the torrent below dashes wildly, filling the valley with its constant roar; suddenly the valley expands again, and on the hill you see the Vetti farm, where the tourist may tarry for the night.

'From the house a zig-zag path leads to the heights above into the deep chasm, from whose edge, by lying flat on the ground, one may venture to look into the depths below, and follow the fall. Another path leads into the valley, and to the foot of Vetti-Foss, or Moerk-Foss. This beautiful waterfall is formed by a stream from two small lakes at the base of the Koldedal plateau 6,510 feet high. From a dark perpendicular wall, forming almost a semi-circle, the stream plunges down from a height of more than a thousand feet. Towards the end of summer, so small is the volume of water that it falls gently in a transparent column of spray, looking the more white by contrast with the dark wall which forms the background. I wondered that this cloud of spray could make such a volume of water, rushing so violently among the rocks that it was with difficulty that I crossed to the opposite bank, from which a better view of the fall is obtained. The soil and rocks are covered with a dark fungus, and everything contributed to make the spray appear whiter. I could see no land beyond, and only a few birch trees on the ridge. As the fall is vertical, only a small portion of the water strikes upon the rocky walls. As I looked the column of spray began to move to and fro as the rising breeze swept round the walls, until it swung like the pendulum of a clock over a space of 250 feet; then came a strong gust of wind, and the whole mass spread into a transparent sheet of spray from top to bottom; as it became still it contracted once more into a white column. For a long time I stood watching the fascinating spectacle, and could hardly tear myself away. It resembles in this changing column of spray the Staubbach fall in the valley of Lauterbrunnen in Switzerland, and still more, according to descriptions and photographs, the upper portion of the Yosemite fall in the famous valley of California. This latter plunges vertically about a thousand feet over a granite precipice, varying much in appearance according to the volume of water in different seasons, and its column of spray, in the same manner, is the delicate plaything of the winds. But the Vetti-Foss has more water. These bridal-veil water-falls are counted by hundreds in Norway.

'If the tourist ascend the grand fiord of the Aurland, on the southern shore of the Sogne, some ten miles west from Laerdalsoeren, and cross the mountain ridge separating this basin from that of the Hardanger fiord, the lovely basin further to the south already spoken of, he will pass by the way the Tvinde-Foss, which pours over a sparsely wooded ledge, three or four hundred feet in height, and its cascades, if not grand, are among the loveliest in Norway.'

But we have not yet done with the Sogne fiord. 'One of the striking characteristics of the Sogne fiord is the varying depth of the water. South of Yttre Sulen, the island lying in the mouth of the fiord, the depth of the sea is stated by Du Chaillu to be about 600 feet; farther inland, between Big Store Hilleoe and Stevsundoe, it is 1,584 feet; a little farther up it diminishes to 1,200 and 900 feet, and immediately south of Poe church it attains the enormous depth of 3,980 feet [more than three-fourths of an English mile]; north of Arnfjord church 3,222 feet; at the entrance of the Aurland, 3,766 feet; and just south of Kaupanger, 2,964 feet [still, more than half a mile, 2,640 feet, in depth]. The branch fiords are much narrower, but their depth of water is also very great. The Sogndal at its entrance, which is narrow, is 132 feet, but about midway it is 1,194 feet, thence becoming near its end 216 feet deep. The Lyster is at its entrance 2,170 feet [half a mile] deep; half-way, 1,176 feet; towards its end 276 feet. Even in the Aardal and the Laerdal, which form the upper end of the Sogne, the sea in the former is 840 feet, and in the latter 780 feet deep.'

In these varying depths may be found indications of an action of glaciers which well deserves study.

Existing glaciers may be considered remains of a sheet of ice or snow, which, in the glacial era, and long after, covered extensively Northern Europe. Glaciers are now, and probably were then, in a state of continuous flux, flowing from a higher to a lower level, as does water, as does tar, as does honey, and as do many substances more tenacious than are they, whenever they are allowed so to gravitate.

It may be asked how can a solid body like ice flow? And the answer is forthcoming. All matter, even the most solid and compact, is composed of minute particles of the substance kept together by mutual attraction, but not in actual contact. When the attracting force can only act within a very limited distance, though powerfully within that distance, the body is friable, easily broken, it may be easily shattered. Thus is it with glass, with sealing wax, with cast-iron; but there are also substances which, when warm, can be spread out in sheets, or drawn out in threads—the attracting force still keeping the particles together in one mass, though individually to some extent dissevered; and even in their solid state they may be found to be within certain limits elastic, allowing of distension without destruction of the attracting power, which brings them into position again when the pressure by which the body may have been bent has been

MECHANICAL ACTION OF GLACIERS.

withdrawn; and ice is elastic, as elastic, apparently, as is glass. But there is also another phase of the same phenomenon which presents itself. It is alleged, and it has been satisfactorily demonstrated, that in the course of passage through a narrow strait or over a steep precipice, a sheet of ice becomes broken up into an infinite number of small pieces, admitting thus of an easy passage of the mass; and that these become frozen together again on their escape from the pressure. And thus, like the tenacious substances which have been named, honey and tar, the ice flows on in an apparently solid, as do these in a semifluid form or consistence.

But in doing so the friction on the bottom and sides of the channel is great. In many places—I had almost said in all countries in these northern latitudes, including our own—there may be seen hard rocks on mountain sides, and in some cases on mountain tops, marked with *striae*, fine parallel hairstrokes, which are attributed to the passage of ice in a state of flux. Besides these there are found everywhere what are called boulders—large masses of rock, which have been torn from sides of mountains by passing glaciers, borne along by the moving stream, embodied in it it may be, and deposited where the ice melting could no longer sustain it. On the surface of the glacier, moreover, there are often seen longitudinal streaks of *débris* which have fallen upon it from higher situated mountain sides as the glacier passed. Moraines, linear deposits of stones and rubbish across valleys, are the produce of such, carried down to the lower edge of the glacier, and dropt as this melted away through the heat. In some valleys there are a succession of such moraines, separated by greater or less distances. These indicate what had at successive periods been the extremity of a glacier previously existing there, which extremity in existing glaciers may be shown to have alternately receded and advanced, and again, it may be, receded and again advanced, only again to recede, after more or less protracted periods of stationary limit, according as the local temperature may have risen, or fallen, or remained stationary for a time.

While on its progress from the higher-lying resting-place to the lower-lying ocean bed, the base of the bed, as well as its sides, is often made to feel the effects of its passage. As the waterfall washes out at its base a basin into which it falls, so does the glacier, not at its extreme edge, but wherever it descends from a higher to a lower altitude, excavate a hollow. And thus may these profound depths in the fiords have been produced.

In a volume entitled *The Forest Lands and Forest Management of Finland* * I have had occasion to remark:—
·With ice as with water, notwithstanding its hardness and its tenacity, it seeks the lowest level to which it can attain; and the glacier is ever in a state of flux from the land towards the lower level of the sea, on its advance grinding away, smoothing, and striating the surface of the rocks, past which, or over which, it flows. The pressure, and consequent abraiding power of a glacier must be tremendous: the *vis a tergo* being such that it treats as mere pebbles in its path ridges, and even hills of considerable elevation, and it seems to pass as easily over them as a deep river flows over the stones that may be in its channel.

Thus may be accounted for the numerous lakes existing in Finland, giving to it its character and its poetic designation The Land of a Thousand Lakes, and the existence of the lakes so abounding in Norway and in Sweden.

In the *Quarterly Journal of the Geological Society*, vol. xviii., p. 185, in a paper by Professor (now Sir Andrew) Ramsay, entitled ' The Physical Geology and Geography of

---

* *Finland: Its Forests and Forest Management.* In this volume is supplied information in regard to the lakes and rivers of Finland, known as the *Land of a Thousand Lakes*, and as the *Last-born Daughter of the Sea.* In regard to its Physical Geography, including notices of the contour of the country, its geological formations, and indication of glacial action, its flora, fauna, and climate; and in regard to its Forest Economy, embracing a discussion of the advantages and disadvantages of *Svedjande*, the *Sartage* of France, and the *Koomaree* of India Details of the development of Modern Forest Economy in Finland, with notices of its School of Forestry, of its forests and forest trees, of the disposal of its forest products, of its legislation, literature, and forestry.

Great Britain,' may be found the first suggestion, and illustration, and proof of this fact. There ' he has shown that the innumerable rock-enclosed basins of the Northern Hemisphere do not lie in gaping fissures, produced by underground disturbance, nor in areas of special subsidence, nor in synclinal folds of the strata, but that they are true hollows of erosion.'

I cite the statement of Professor Geikie, in his *Scenery of Scotland, viewed in connection with its Physical Geography ;* and to this work I am indebted for the following illustrations:—

' Lakes, at least those which mottle the surface of Scotland, may be grouped into three classes : 1st, those which lie in original hollows of the superficial drifts ; 2d, those which have been formed by a bar of drift across a valley or depression ; 3d, those which lie in a basin-shaped cavity of solid rock.'

Lakes of each of these kinds may be seen in Scandinavia and Finland. It is in regard to the formation of the last description of lakes that there is any difficulty—the formation of a cup-like hollow in solid rocks, sometimes along the line of a valley, sometimes on a plateau, sometimes on a hill top, or on a watershed.

There are in many rivers deep holes. At the Cape of Good Hope one hears constantly of *See-Koo vleys*, or hippopotamus holes, and occasionally, even in the rocky bed of a river, we find cylindrical cavities called pot-holes. In the bottom of such are generally found a few well-rounded pebbles and boulders. The cavities are due to the circular movements of these or other stones and boulders, which, caught by an eddy, have been kept whirling there, and by friction abrading the rock they have gradually formed, these holes working downward into the solid rock. And often on the sea-shore may be seen cavities lined with sea-weed and filled with sea-water, each a natural aquarium. Some of these are formed, as are the pot-holes, by boulders lying in their bottom which have been kept whirling round in the eddies of a vexed tideway instead of a rapid brook or river.

But it is not thus that these rock lakes have been formed. Of the theory of Professor Ramsay the following illustration is supplied by Professor Geikie:—' A river of ice is not bound by the same restraints as those which determine the action of a river of water. When a glacier is, as it were, choked by the narrowing of its valley, the ice actually rises. In such places there is necessarily an enormous amount of pressure, the ice is broken into yawning crevices, and the solid rocks suffer a proportionate abrasion. The increased thickness of the mass of ice at these points must augment the vertical pressure, and give rise to a greater scooping of the bed of the glacier. If this state of things lasts, it is plain that a hollow or basin will be here ground out of the rock, and that once formed, there will always be a tendency to preserve it during the general lowering of the bottom of the valley. On the retreat of the ice, owing to climatal changes, this hollow, unless previously choked up with sand and stones, will be filled with water, and form a lake. It will be a true rock basin, with ice-worn surfaces around its lip, and over its sides and bottom.

'And such is the appearance presented by many a lake and tarn in the Highlands of Scotland. One of the largest and noblest of the whole—Loch Awe—may be taken as an illustrative example.'

Nor is it only the formation of single lakes which can be thus accounted for; a continuous succession of lakes in the direction of the movement of the glacier may be thus produced.

As popular illustrations of the mode of operation I may cite the following:—Young boys, and girls too, amuse themselves making what they call 'ducks and drakes' by throwing flat stones across a placid sheet of water, as nearly parallel to the surface of it as they can, causing them to skim along and above the water, touching and rising again and again, rebounding in ever-diminishing bounds till they sink. The same phenomenon may be seen on a larger scale in the recochetting of a cannon ball fired at a target out at sea; and the same thing may be seen in the effects of the wind striking the surface of the water in a river, in a lake, or in the sea, for rarely, if ever, does it blow horizontally or parallel to the surface of the water.

In another volume, entitled *Forests and Moisture*,* I have had occasion to refer to another and different phenomenon occasioned in the same way. It is of frequent occurrence at the Cape of Good Hope, and in local phrase is spoken of as the Devil's Table-Cloth on Table Mountain. At these times the summit of the mountain is covered with a dense mass of beautiful white fleecy cloud in constant flow over the precipice, and pouring down the almost vertical front of the mountain facing Table Bay as if threatening to bury in an avalanche the capital of the colony at its base; but long ere it reaches the town, notwithstanding the continuous flow, it stops; to that line it flows on continuously, but beyond it not; there the cloud, in unceasing flow, terminates, the spectator sees not why.

The beautiful and interesting phenomenon is occasioned by a south-east wind, which up to the Table Mountain range, was undimmed. The wind was strong, but the sky blue and serene, though the wind was loaded with vapour —vapour dissolved and invisible.

But, passing over Table Mountain, the elevation of this is such that the decrease of temperature, consequent on expansion under diminished pressure, bringing this below the dew-point, the moisture is deposited by the air in the form of a cloud, which, as it reaches, at a lower level to leeward, a locality with a higher temperature, the moisture is

---

* *Forests and Moisture*: or, Effects of Forests on Humidity of Climate. In which are given details of phenomena of vegetation on which the meteorological effects of forests affecting the humidity of climate depend,—of the effects of forests on the humidity of the atmosphere, on the humidity of the ground, on marshes, on the moisture of a wide expanse of country, on the local rainfall, and on rivers,—and of the correspondence between the distribution of the rainfall and of forests,—the measure of correspondence between the rainfall and that of forests,—the distribution of the rainfall dependent on geographical position determined by the contour of a country,—the distribution of forests affected by the distribution of the rainfall,—and the local effects of forests on the distribution of the rainfall within the forest district.—Edinburgh: Oliver and Boyd. London: Simpkin, Marshall, & Co. 1877.

again absorbed, and the air loaded with it is again transparent, as is all the air around, and as it was itself before passing over Table Mountain in its course.

From Claremont, or Wynberg, or the Flats, or any place to the back of Table Mountain, it may be seen that the cloud is not blown to the mountain, but that there it first appears; and if some few cloudlets, formed over the crests of hills belonging to the range situated to windward, be seen sailing towards it, it is evident that they are 'A sailing, a sailing with the wind,' and not attracted only, for none are seen floating toward the Table-Cloth in other direction than that in which the wind blows.

Of this phenomenon Sir John Herschel writes, ' That the mere self-expansion of the ascending air is sufficient to cause precipitation of vapour, when abundant, is rendered matter of ocular demonstration in that very striking phenomenon so common at the Cape of Good Hope, where the south or south-easterly wind which sweeps over the Southern Ocean, impinging on the long range of rocks which terminate in the Table Mountain, is thrown up by them, makes a clean sweep over the flat table-land which forms the summit of that mountain (about 3,850 feet high), and thence plunges down with the violence of a cataract, clinging close to the mural precipices that form a kind of background to Capetown, which it fills with dust and uproar. A perfectly cloudless sky meanwhile prevails over the town, the sea, and the level country, but the mountain is covered with a dense white cloud, reaching to no great height above its summit, and quite level, which, though evidently swept along by the wind, and hurried furiously over the edge of the precipice, dissolves and completely disappears on a definite level, suggesting the idea (whence it derives its name) of a "Table-Cloth." Occasionally, when the wind is very violent, a ripple is formed on the ærial current, which, by a sort of rebound in the hollow of the amphitheatre in which Capetown stands, is again thrown up, just over the edge of the sea, vertically over the jetty—where we have stood for hours watching a small white cloud in the zenith, a few acres in extent, in violent internal agitation (from the hurricanes of wind blowing through it), yet immovable as if fixed by some spell, the material ever changing, the form and aspect unvarying. The " Table-Cloth " is formed also at the commencement of a "north-wester," but its fringes then descend on the opposite side of the mountain, which is no less precipitous.'

Other illustrations, perhaps more pertinent, are supplied by sand ripples on the shore, and by the contour of sand drifts, while an illustration of reboundings out at sea, like to the ærial rebound described in the passage cited from the writings of Sir John Herschel, are supplied by banks in some of the Argyleshire lochs, vertical to the line of descent of the Highland glen down which in pre-Adamic times poured the glacier which hollowed out the basin. The confining sides of a valley once formed would elongate the furrow or depression thus created in a direct rather than a cross direction; but the alternate elevations and depressions, and thus the succession of pits in the *thalweg* of the glacier, may also have been thus produced.

In view of this it becomes more easy to see how pools of such depths of water as 780 feet, 840 feet, 1,584 feet, 2,964 feet, 3,766 feet, and 3,980 feet, may have been produced in the Sogne fiord and its branches, while the depth of the sea at the mouth of the fiord is only 600 feet, may have been produced, and successions of such hollows in the line of the fiord and lateral branches. In a like way may have been produced the basins of such lakes in the interior of the country as the Miosen with a depth of basin 1,110 feet below the level of the sea, corresponding to the depth of the sea basin in the outer portion of the Christiania fiord.

The force with which the water of such falls as have been described impinge on the basin at their base must be tremendous; but water is a liquid yielding material. Imagine what must have been the impinging force of an ice-fall from such heights as those of some of the mountains around, 6,500 feet, and 7,620 feet—a mile and a-half above the level of the sea—into a basin 3,980 feet below that level, a fall of two miles and a quarter (Can it be !), and there seems nothing difficult in further imagining it excavating such basins as have been mentioned.

Reference has been made to Loch Awe. The western lochs of Scotland appear to abound in such pools, and in chains of them.

In Gairloch, there is at the head of the loch, I was informed when there lately, a pool very much deeper than the basin of the loch, and the basin of the Clyde beyond; separated from this by a low range of high land, on which is situated Shandon, Row, and Helensburgh, is a dry valley of much greater depth in its upper and middle stretches than at its lower extremity. And a little beyond is Loch Lomond, with a deep pool towards its upper extremity; and another of less depth in the line of its basin a little below and within sight of Inversnaid.

In accordance with what has been said in regard to the *striae* to be seen on the face of the rocks, two theories have been advanced in regard to the production of such *striae* observed elsewhere.

Both theories attribute them to the action of ice. In one of these, advanced and expounded in a volume entitled *Frost and Fire*, by Mr J. F. Campbell, they are attributed to the friction of icebergs and icefloes, drifting from the north on ocean currents. In the other, advanced, maintained, and illustrated by Agassiz, Ramsay, Lyell, Chambers, Jamieson, and Geikie, they are attributed to the grating action of glaciers, or land ice, formed where they are seen, or at a somewhat higher level, and continuously descending in a state of flux to a lower level. Thus does it appear to have been here.

Professor Esmark in a paper *On the Geological History of the Earth*, in *Jameson's Journal*, October 1826 to April 1827 (p. 120), describing certain phenomena near the embouchure of the Sogne fiord, says:—'On this rock there seemed to me to be proofs of the powerful operation of ice. I found that the precipices on the side of the mountain next the sound were several [hundred?] feet in height, and perfectly perpendicular, and though they were composed of boulders cemented together, they were perfectly even and smooth. If these precipices had been the effect of rents, attended with successive masses tumbling down, then the boulders adjoining the rent must have been found adhering, sometimes to the one and sometimes to the other of the separated masses (those which have fallen into the sea are no more to be seen), and in that case the boulders left on one mass must have left a mark of themselves in the corresponding one. This, however, was by no means the case, as the rock which remained was perfectly smooth, and had the appearance as if these boulders had been cut across with a sharp knife. I can explain this phenomenon in no other way than by supposing that large masses of ice, pressing through the sound, have cut these precipices lying parallel in the direction of the sound.'

Forbes, citing this statement, shows that the reference is to the action of glaciers, and not to that of floating ice.

Principal Forbes, who had gone to Norway to study the glaciers of that country, writing of the locality below what has just been described, says:—'In the course of the forenoon we passed the opening on the great Sogne fiord, the most ramified in Norway, stretching landward not less than 110 English miles, to the head of the Lyster fiord, one of its farthest tributaries. Having heard much of the surprising and gloomy cliffs of the Sogne fiord, I was disappointed to find its entrance tame, undulating, and without much interest, whilst the higher mountains were too remote, or too much concealed by the intermediate hills, to produce a favourable effect. The character of the rocks and islets of the fiord was, however, worthy of notice, though far from picturesque. They present to an excessive degree the forms of *roches moutonnées*—the bare grassless surfaces, dome-like, or undulating in tedious monotony, so characteristic of glacial action, with the usual accompaniments of flutings and polished channels. The material of the rocks renders these impressions of external friction still more striking, for it is chiefly a coarse conglomerate, of which every part, the boulders as well as the cement is cut as by the lapidary's wheel. The wonderful extent over which these appearances occur, and the unsparing severity with which the natural inequalities of the most obdurate rocks have been smoothed down, is strikingly impressive, when we couple it with the fact that if glaciers really were once much wider spread than at present, this vast chasm was the natural outlet of an icy flood, drawn from a more extensive origin than any other existing in the north of Europe.'

There may seem to be here an account of the general appearance presented by the fiord at its mouth differing from that given by the more graphic pen of Du Chaillu. I look upon the latter as probably the more valuable as an account of its picturesque effects; that of Forbes as more valuable as testimony of a scientific student of distinguished attainments, giving his special attention as he had been doing long, to the indications of glacial action. On this point both are agreed, and Forbes speaks only of the mouth of the fiord, Du Chaillu takes us into its recesses.

By some of the geologists whom I have cited it is held that in what is known as the glacial period Scotland must have been covered with one wide-spread sheet of ice and snow of great thickness, as at the present day is Northern Greenland, where there may be seen an interminable glacier extending league upon league, broken only by some black hill top or mountain peak that rises as an island above the sea of ice. But there this vast sheet is ever, even while being replenished by fresh falls of snow, slowly and persistently flowing, or rather creeping, down to the sea,

MECHANICAL ACTION OF GLACIERS.

covering the face of the country, filling up the valleys, mounting over the hills, and pressing with constant resistless force upon all the rocks over which it advances; and blocks of stone, either loosened from the mountain by frost, or torn off by the moving glacier, are jammed in in the rear, and pressed along the rocky bed or sides of the valley; and the stones, mud, gravel, and sand thus borne along act like files, scratching and scoring the hardest rocks, and being themselves scratched by the same process. As it is now in North Greenland so must it have been during the glacial period in Scotland. There we find the rounded, filed-down projections on the mountain top and on the mountain sides, and the parallel *striae*. So must it have been at the same period, and on to a later time, in Norway; and thus many numerous phenomena presented by the mountains and the rocks there be satisfactorily accounted for.

The rasping of the ice, charged with fragments of stone, and gravel, and sand, would occasion *striae* and markings on the rocks, and by the direction of these may be traced the direction of the movement, while variations in the direction of these can be accounted for.

The *striae* produced by glaciers are generally apparently parallel and straight. The normal aerial currents, popularly known as the 'trade winds,' produced by well-known causes, follow a curved direction, throwing off eddies both upward and horizontal. Similar currents and eddies have been observed in the ocean. Like eddies may be seen in the river, and even in the cup of tea, produced by upward currents from the dissolving sugar; and *striae* may be seen following a curve more or less expanding, and more or less contracting, and variations in their direction may have been similarly produced.

Another of the results of the flux of a glacier is the formation of a deposit of stones at the extreme edge of it; stones which have been borne along on its surface, or, it may be, in some cases a little way beneath this by the slow massive advance of the body of ice, on reaching the extremity of it where it is slowly melting away, though continuously replaced, like the lower fringe of the so-called 'Devil's Table-Cloth' on Table Mountain, which has been spoken of, there drop and accumulate. We have many indications of such glacial action in Norway.

It requires sometimes an experienced geologist to judge satisfactorily in regard to such deposits, and determine whether they be the products of glacial or of torrential action. Several discussions relative to the origin of large deposits in France are cited by me in a volume entitled *Reboisement in France** (pp. 101-117, &c.) In regard to the deposits in Finland there is little room for reasonable doubt that they are moraines, and not what in France are designated *lis de dejection* from torrents.

Norway abounds in similar indications of glacial action in the ages to which these are referred, and many tourists have recorded their impressions of the appearances presented by the glaciers and snow-fields which still exist.

The indications of such glacial action—carrying off boulders and stones from mountain tops and mountain sides, transporting them to the extremity of the glacier, however remote, and depositing them there—are to be found everywhere. But the indications of this action, according to the authority I have cited, are various. They consist of *striae*, or somewhat parallel markings, on the surfaces of the rocks, of moraines, or heaps of stones and gravels, of erratic deposits, of beds of clay, and of *débris* of marine shells. The continental ice, at the time of some of these deposits, must have covered the whole Scandinavian peninsula. And it is supposed that

---

* *Reboisement in France;* or, Records of the Re-planting of the Alps, the Cevennes, and the Pyrenees with trees, herbage, and bush, with a view to arresting and preventing the destructive consequences of torrents: in which are given a *résume* of Surell's study of Alpine torrents, and of the literature of France relative to Alpine torrents, and remedial measures which have been proposed for adoption to prevent the disastrous consequences following from them,—translations of documents and enactments, showing what legislative and executive measures have been taken by the Government of France in connection with *réboisement* as a remedial application against destructive torrents,—and details in regard to the past, present, and prospective aspects of the work.—London : Henry S. King & Co. 1876.

MECHANICAL ACTION OF GLACIERS.

at this time the Baltic communicated with the White Sea.

It is alleged that it was after the melting of the continental ice, and while the glaciers were thus receding from their extension to the valleys, to the summits and elevated ridges, where they still maintain their position, that the transition issuing in the existing state of the country occurred. At about the same period the communications between the Baltic and the White Sea were interrupted, and the ground of the Scandinavian peninsula, in the southern part of Norway, was elevated about 160 metres, or 530 feet. Old shore lines, banks of marine shells, and sand terraces, supply indications, both in Norway and Sweden, of this having been the case, and prove at the same time that the elevation varied with time; and that during a lengthened period it was a time of stagnation, perhaps even of subsidence. In Norway there exist banks of marine shells at, at least, two different altitudes, the one about 150 metres, or about 500 feet, the other about 300 metres or 1000 feet, above the present level of the sea. With the origin of these heaps of shells we have not at present to deal; it is their position alone which here concerns us. The higher-lying banks contain Arctic shells; the lower-lying ones northern shells, corresponding to the species which live at the present day along the coast. Considering both to have been collected on the shore of the ocean —the upper heaps in days more remote, the lower ones in days nearer to our own, these heaps supply indications at once of an elevation of the land, and of the other phenomena referred to in connection therewith. In our own time the coasts of Sweden, on the Baltic, are supposed to be rising at the rate of a metre, or forty inches, in a hundred years.

Old moraines exist at a succession of altitudes up to that of those which are being formed under our eyes— sometimes lying across the depth of the valley, terminal moraines; sometimes along the line of the water-courses, lateral moraines. Large moraines exist on the two sides of

Christiania fiord, the one in an almost direct line from Moss to Fredrickshald, the other in a line almost direct from Horten to a little below Larvick; the two form nearly equal angles with the direction of the Christiania fiord. Their lengths, about 45 kilometres, are also pretty nearly the same. But the *striae* produced by erosion show that these are two separate terminal moraines. The eastern moraine was formed by the continental ice coming from what is now the forest land of the Folloberge and Smalene; the second has been formed by ice coming from the Skrimsfjelde. The great roads of the kingdom traverse these ancient moraines, which supply excellent material for the construction of them.

Small moraines may be seen in a line almost direct between Droebak and the south end of the Lake Oieren, at the end of Mandalsvand, near Christiania, at five kilometres from Christiania, across the railway to the Miosen lake, in some places through the valley of Lier; at the waterfall of Vestfoss, in Eker, fifteen kilometres beyond Drammen, on the low lands of Jæderen, and near to many of the West fiords. Some banks of sand and rounded stones have a great resemblance to moraines, but are not such. They have been deposited, on the contrary, under the surface of the sea; the round stones do not present themselves pell-mell, as stones do in moraines, but in regular beds. One of these connects the Drammen fiord with Svelvik; higher up, near the Dramselv, a similar bank may be seen of more strange outline, and still higher, between Drammen and the waterfall of Vestfoss, the Rygkollen forms a third.

The *débris* of moraines of the glacial period often cover great areas on the mountains. Erratic deposits, and transported blocks, both great and small, are scattered everywhere high and low on the summits of very high mountains, but not, however, on the highest. Thus they are found on many mountain summits on the borders of the Jotunfjelde, in Valders, and in Grunbrandsdal, but not on the highest of these mountains. These erratic deposits are also found in the large moraines of which mention has been made.

Débris of moraines cover likewise great extents of low-lying lands and valleys; and they there entail no small amount of labour in clearing the ground for cultivation. To obtain ground fit for cultivation 50 centimetres, or 20 inches, in depth, it is often necessary to dig out and carry away erratic stones which would have covered the whole area to a depth of a metre, or 40 inches.

Beds of clay and banks of shells of the glacial period are often found spread over areas of great extent, especially in districts open to deposits from Silurian regions of limestone and argilaceous schists, which have been ground down and worn away during the glacial period. Transported as moraine mud by the waters flowing from the glaciers, they have formed beds under the surface of the sea in these ancient times. In different places the lower bed of clay is limestone or marly clay; often it is filled with diversely shaped lumps of hard marl, which sometimes enclose fossils. This marly clay has made fertile the flat country of Remerike and of Smaalehnene, the east part of the valley of Christiania, Eker, the west flat part of Ringerike, Jæderen, and the flat country of Drontheim. The marshy and peat lands of Listerland and of Jæderen rest on a bed of marine clay. For all of this information I am indebted to the reports by Dr Broch and Principal Forbes; and this is also the case with what follows. It is only the circumstance of my having neglected to mark quotations in MS. notes made years ago, which prevents me from inlicating such here.

By the fragmentary shells found in cretaceous sand, the succession of molluscs, both marine and those of fresh-waters, may be traced. For our present purpose it is enough to notice that by the remains of these the ground has been enriched.

In going up one of the great water-courses from the sea towards its source we come first to a level where the ancient marine formations terminate, and where the continental formations begin. This point is passed in the valley of the Glommen at Kongsvinger, and it is equally marked in numerous other localities. At a higher altitude in the valleys we come to a second slip, formed by the rocks contracting the valley, or by a moraine. Behind or beyond this slip there is found generally the basin of the valley covered with sand, or near this more rarely with clay; and so on, as further and further the valley is ascended. In proportion as the continental ice was thick, some valleys have remained a shorter or a longer period filled with ice. The lower limits of glaciers have always been receding to higher altitudes, and in places where they have remained stationary for a long time they have left moraines. These have there formed dykes, behind which was thus created a reservoir in which the clay carried off by the waters fell to be deposited in layers. Beds of clay in extensive areas above the ancient sea level are rarely met with; but they are met with in certain places behind these dykes, amongst the filling up of the later formed slips in the basin of the valley.

In the valley of the Glommen, in the water-course of the Guldbrandslagen, and in the basin of the Dokka, all of these may be seen

On the west coast of Norway the same succession of slips may be seen, but with a character somewhat different on account of the shortness and the depth of the valleys. At the extremity of the interior ramifications of the fiords there rises often a rampart of gravel and erratic deposits; it is the moraine of the glacial period. Behind this there is ordinarily found a lake at a height a little above the level of the sea, while the depth of the lake is often considerably below the sea level. Such lakes are numerous, and beyond them, landwards, there are met with on many points a distinct slip in sandy ground, steep towards the sea, but flat towards the interior of the country; it marks the ancient level of the sea.

In the diocese of Drontheim, where the valleys stretch

out to a considerable length, and are not so steep, we find the country rising in steps corresponding to those of Southern Norway. Here considerable spaces are found above the ancient level of the sea. In Orkedal the first step which marks that ancient level is a little to the south of the church of Medal; in Guldal it is at the church of Stœren.

CHAPTER XVI.

APPEARANCES PRESENTED BY GLACIERS AND SNOW-FIELDS.

OF the Justedal glacier, adjacent to the Sogne fiord, Du Chaillu writes:—'This field of snow, the largest in Scandinavia, covers a continuous tract of over eighty-two English square miles, its depth in many places reaching 1000 feet. It comprises the area bounded on the north by the North fiord, on the south by the Sogne, on the east by the Justedal valley, and on the west by the Sogne fiord. Its lower part is entirely fringed by glaciers which flow in every direction. The glaciers in the Fjaerland fiord are three miles inland; the extremity of the Poyum being about 400, and the Suphelle 160 feet above the sea. The backbone or rocky ridge of this mass of snow has an average height of 5000 feet, the highest point lying between Stryn and Justedal valley, Dalskaupos peak reaching a height of 6,410 feet in the eastern, and 6,110 in the southern part.

'At the head of Gaupe fiord, on the Lyster, is the valley of the Justedal, which derives its name from the great glacier which overtops its mountains. At the entrance is the hamlet of Roeneid, with a comfortable inn, where horses can be procured. A narrow road, used as a bridle path, and passable with a cariole for a distance of six or seven miles, leads to the end of the valley.

'About fourteen miles from Roeneid stands the plain parish church of the valley, surrounded by a rough stone wall and the humble churchyard, with only a few wooden crosses; the adjacent parsonage has a small garden, and a few patches of barley and potatoes, and may be said to be the only clean and comfortable place in the vicinity.

'A few miles farther on is the Berset glacier, the first in the valley, and near it is the poor hamlet of Nygaard. From the deep-blue cavern at the base of this glacier flowed with great force a dirty stream into the valley, and close to the icy edge was a parallel line of boulders, stones, and sand, left behind by the retiring mass. Beyond this were several other transverse ridges, formed by similar deposits, showing that the glacier is fast retiring.

'After a pony ride of twenty-eight miles I came to Faaberg, the last hamlet of the valley, containing several well-stocked farms, and surrounded by verdant fields and meadows. . . . .

'From Faaberg the path was extremely rugged. The ceaseless noise of the rushing water, formed chiefly by the glaciers of Bjoernestig, Lodal, and Stegeholt, at times was so great as to drown the voice. . . . Winding our way for a while through meadows and woods, we saw in the distance, at the end of the valley, the Stegeholt and Lodal glaciers; the summit of the peak is 6,410 feet above the sea. At the end of that wild valley was the usual moraine with rounded stones, pebbles, and sand, left by the retiring glaciers. The streams from them divide and meet again; the current was very strong, and the water so dirty that our horses were almost afraid to cross. One would naturally think, not knowing the laws which govern the movement of a glacier, that a stream created by the melting of pure ice could only produce the clearest water; on the contrary, the very nature of a glacier prevents any other sort of stream. In June, and even in the beginning of July, these streams are unfordable. The Lodal glacier was covered with dirt, stones, and *débris* from the mountain side. Its cavern was by far the finest and longest that I had seen, being about 20 feet wide; from it a turbid river rushed with great force. The beauty of this cavern cannot be adequately described, the blue colour of the ice gradually became deeper, finally merging into an intense inky-blue. Owing to the great pressure every air-bubble had been expelled, and the whole mass was clear and transparent; the cavern appeared like a tunnel cut through a mountain of sapphire. Unfortunately I could not explore it on account of the great depth and velocity of the water, as it ran between two stone ridges, split by the ice. The retiring glacier had uncovered part of a spur or hill of gneiss, which had obstructed its march, and which was split into several enormous parts, which were still in contact with each other. A considerable number of boulders were resting on the frozen mass, some supported on pillars of ice, which were prevented from melting by the protecting shade of the stones. In places the glacier was white, not from snow, but in consequence of the cracking of its surface and numerous air-cells. It was easy to see that the Lodal had formerly been much lower down the valley, and that the transverse glaciers we had met on the way were once its lateral branches, the whole forming a single vast frozen river reaching the sea, retiring, advancing, again retiring. Thus the ice ground deeper and deeper into the rocks; the same marks were visible, left by that which had retired the year before. I heard a rumbling sound, and had hardly raised my eyes when a huge stone from the glacier rolled within a few feet of me; and I had hardly seated myself the second time when I saw another stone roll down carrying with it in its flight several lesser ones.'

M. du Chaillu goes on to say: 'A glacier is not an immovable mass closely attached to the mountains, but a body slowly impelled forward by the immense pressure of the upper portions. On its way the mass slides down grinding its rocky bed, thus deepening and enlarging its channel, day by day; its silent power, overcoming all obstacles, carries with it whatever has been buried in the icy stream, such as stones that have fallen from the mountain sides, earth, and sand, which combine to render the water turbid, and to form the moraines. It has the character of a stream; it is a moving river of ice fed from the *sneebraer* or perpetual snow-fields above, modifying or creating its channel, eroding valleys, often covering vast areas, an agent of great destructive power.

'The motion of a glacier being largely due to expansion from the consequences of its melting, is slower at night than during the day, and in winter than in summer; the movement is greater in the middle, than on the sides, where it is held in check by friction, and also more sluggish at the bottom than at the top. A glacier will accommodate itself to the sinuosities and unevenness of its bed, expanding or contracting like the waters of a river, and will precipitate itself over a ledge, making a cascade of ice: these I have seen in almost every glacier in Norway. The ice is often broken transversely, the moraines are engulphed in the crevasses and lost. The main glacial stream starts with a moraine on each side; long dark bands raised above the ice are formed by the stones and earth which have fallen down the side of the mountain, in the same manner as the heaps of stones and *débris* we find at the base of mountains, and in many ravines and valleys. These lateral or marginal moraines vary in height according to the amount of the deposits massed together, and to the time of their formation; they range from a few feet to twenty feet in height, but are never much more, for there is no time for accumulation; the material is collected as the ice moves downward, and the motion of the Norwegian glacier may be a few hundred feet a year. These moraines stand in regular ridges, and are slowly and surely carried to the end of the glacier; their origin, by the materials, can often be traced back for great distances. As the frozen river moves onward, it is joined by others, all uniting in one solid mass; the moraines meet side by side, and remain distinct on the journey down. The number of these moraines indicates how many branch streams have united with the main trunk. Sometimes a glacier is compelled to make its way through a narrow defile; then the mass of ice contracts, and becomes deeper, and a grinding process takes place on the sides and at the base with immense force; many valleys with perpendicular walls have been formed in this manner. Not far from Lodal is the very interesting glacier of Stegeholt, reached by again fording the Lodal river. The end of this glacier is narrow, and the ice comes through a contracted gorge choked with large stones, which prevented me from seeing the terminal cavern.

'On the left bank, to a certain height, birch trees were abundant, and there was a dense growth of grass and weeds within a few yards of the ice. Here also I saw evidence that the ice had much diminished that year. Numerous large boulders, forming longitudinal moraines, were stranded along its sides. The crevasses indicated a powerful strain; through the cracks, which crossed the whole breadth of glacier, you could see the deep blue colour growing darker and darker with the increasing depth.

'We have given a description of retiring glaciers. We must also speak of those which are advancing with an irresistible power.'

Following an account of the Ringedal waterfall, our traveller writes:—'A row of one hour on the fiord brought us to Odde; from which the tourist should not fail to visit the Buer-brae-en, one of the glaciers, of the Folgefonn. A ridge of mountains crosses the Folgefonn, in a north-easterly direction, forming the Svartdal—black dale—and the Blaadal—blue dale; and another ridge forms the Kvitnaadal. Blocks of stone mixed with sand showed their unmistakeable origin. The glacier had reached this point years before, had retired, and was now again advancing; while higher up, our path continued through a wood, in which numerous moss-covered stones could be seen showing that the glacier had not reached that altitude for a very long time.

'The view of that narrow glacier was imposing, impressing the mind with a sense of the great power of destruction possessed by a vast body of moving ice. In the study of other glaciers which were retiring, we have seen how the boulders and smaller stones have been deposited in the fields in former times, and we could trace, by the marks of the ice on the rocks, the course taken. But now standing before the Buer-brae-en, we could understand how valleys had been dug out of the solid rock by that most destructive form of water the glacier. The huge irresistible mass was still advancing slowly, and had been doing so for a long time. My guide said it had advanced more than fifty feet since the previous year, driving everything before it. All along the base of the ice was a transverse ridge of earth in which fresh greensward and stones were mingled together, which the glacier pushed forward as it glided over the rocks. On the right was a huge mass of rock which had been torn apart by the pressure of the advancing ice. The weight which had overcome this obstacle must have been enormous, for the evidence of such terrific force was before my eyes. Not even the solid mountain walls, composed of the hardest of our rocks, could arrest the forward march of the terrible glacier. This block of granite, torn from the mountain side, was about twenty feet long and fifteen broad. It had been broken unevenly, and was still covered with moss. A part of it was overlapped by the ice; and the upper stratum of the glacier having a stronger current than the lower would finally run over it, and hide it from view as the onward march continued; and when the glacier again retired the boulder would be deposited in some new resting place. The glacier came down a steep gorge leaping three distinct ledges of rock, and it was crowded between solid walls not more than 250 to 300 yards wide towards its end. The moraines seen higher up on each side above were engulphed further down into deep crevasses formed by the pressure of the ice and ledges. On its left were towering mountains; Mount Reina being 5,210 feet above the sea, and the second highest point of the Folgefonn. The ice was of a magnificent blue; the cavern was small, but extremely beautiful; and its stream was far from being as dirty as those of the glaciers of the Justedal. Lower down in the valley, not far from the glacier, was the Buer farm; and from the mountain side came a cascade between 700 and 800 feet in height. The owner of the little farm was in great tribulation. He saw with much anxiety the steady advance of the ice which had already destroyed some of his pasture land at the head of the valley, and in a few years would probably sweep away the little wood which we had passed on our way up; then the farmer would be compelled to find new quarters, and perhaps be a ruined man. He had tried to sell his farm, but nobody was willing to buy it, fearing to cast away their money. It would not be strange indeed if in the course of forty or fifty years this glacier should reach the very shore of the Sandven lake, whence it could go no farther, for the ice would melt in the water; but glaciers are fickle both in their forward and retrograde movements, and in a few years the Buer-brae-en may retire instead of advancing.'

It is interesting in looking upon such scenes —forgetting for the time, if we be there in the prosecution of the study of forest science, the importance of these glaciers as perennial sources of supply of water in the rivers which are made use of for the floatage of timber from remote high-lying lands to the coast —to look upon them as modern illustrations of what was written well nigh 3000 years ago in Ecclesiastes—attributed to the Royal Preacher, who 'spake three thousand proverbs; and his songs were a thousand and four; who spake of trees, from the cedar tree that is in Lebanon, even unto the hyssop that springeth out of the wall; who spake also of beasts, and of fowl, and of creeping things, and of fishes;'—and that twice repeated, 'That which hath been is now; and that which is to be hath already been. . . . 'The thing that hath been is that which shall be; and that which is done is that which shall be done; and there is no new thing under the sun. Is there any thing of which it can be said, see this is new? It hath been already of old time which was before us,' —and to look upon them as supplying us with data for the explanation of phenomena, which abound in these northern lands as elsewhere.

Other travellers have supplied additional sketches which help to fill up the outline.

William's writing of his voyage to Hammerfest, in the far north, says:—'At about four o'clock on the second morning of our return journey we passed some remarkable glaciers near to the Havnes station; one of them very nearly reached the sea. We were near enough to examine them pretty fully, and with the aid of telescopes or opera glasses to look down the blue *crevasses* which rib the lower parts. They exhibit the whole phenomena of glaciers at one glance; there is the snow-field or névé above, the source from which the true glacier is derived; the deep lateral valley narrowing downward, one of the essential conditions of glacier formation; then the ice torrent, with its sharp billows and blue chasms, filling this valley, and carrying with it in its slow descent blocks of rock forming the moraine, which, when deposited at its boundaries, will remain to mark its place, though the climate of the whole region should change, and the ice and snow all melt away.'

That was in the far north. Writing of his visit to the Romsdal in the south, he says:—' Veblungsnaesset is the port of the Romsdal, which valley is the "lion" of all Norway; the Norwegians themselves travel long distances to see it. In the *Christiania Illustrated News* there are numerous wood-cuts of its finest scenic features, and every Englishman who comes to Norway is told that he must see it, and his expectations are raised to the highest.' The descriptions given by him of the waterfalls are numerous, varied, grand, and picturesque, and seem to show that ' the Romsdal can safely bear this terrible ordeal of much repeated praise.' But what concerns us at present is the following statement:—' In nearly all the tracks and hollows of the dark precipitous rocks are patches of snow, some of them so low as almost to touch the corn fields; for amid all this savage sublimity there are rich substantial farms. These farms are due to the table-land of the terraces, of which there are two very distinctly marked. Besides the snow patches there are Lilliputian glaciers in abundance, where the whole history of glacier formation is shown at a glance. There are the snow-fields above filling a basin, from which dark peaks arise; the basin has a downward opening, a notch leading to a little steep trough-like valley that closes in below. In the upper basin the snow surface is thawed by the sun, the water sinks into the spongy snow below, freezes again on its way, and binds it all together as a seeming solid, but capable of yielding to the pressure of the mass above, and the expansion of refreezing; this pressure forces it through the notch of the upper basin into the lower. As it passes over the bend from the lesser to the greater declivity it is split upon its surface by this bending, and the blue crevasses are formed. In squeezing so forcibly through this opening it polishes its rocky sides, and the fragments of stone that are torn away, or that fall upon it, become bedded into the ice, and when they reach the portion that slides upon the rock they groove it with parallel lines, which will mark the place where these glaciers have been if in future ages they should cease to exist.

'There are other snow basins which fail to form true glaciers, owing to the want of the trough-like valley below, that closes in at its lower part. Yet in these there is evidently a downward flow, or advancement of the ice and snow, which is forced through the notch; but this notch communicating with a long straight trough like a water gully, the foremost of the advancing mass bends over until it becomes detached, and then forms an avalanche instead of a glazier. Several of these small avalanches came down during my walk. I mistook the first for a water cascade, until its cessation, and the thundering rumble which followed undeceived me.

'In these I found an explanation of the snow patches nearly level with the corn fields; for each of the avalanches deposits itself as a sort of talus, or sloping delta-shaped heap at about that part of the terrace which must have been the shore of the ancient fiord. All these avalanche tracks are smoothed by the falling snow, and ice, and stones; they are probably scratched and grooved likewise, but this I cannot positively affirm, as they were on the opposite side of the river.'

In writing of his journey in still higher lying lands, whence flow the rivers terminating in the Romsdal fiord, and the Sogne fiord, and several others in the district called collectively the 'Nord fiord,' although there is no individual fiord bearing that name, he says, 'The ascent of the valley towards the snowy wilderness of the Nord fiord and Jostedals Bræen is by an abominable path, over the wreck of glacier moraines, and through thickets of low beech trees, or rather bushes; the elastic arms of which, entangled with each other, continually bar the way, and springing back as they are bent aside, pick off one's hat, flog one's face, and take most tantalising liberties with the knapsack behind.

'The vegetation soon ceased, and I came upon a waste of loose stones, with sloppy snow between, and every vestige of the track obliterated by the thawing.' The difficulties and adventures in which he was thus involved are graphically told. He found himself at the head of a valley terminating in three peaks, the centre one being just in the direction he ought to take, but that was impossible. He must follow one or other of the hollows between; but which? It was evident that these courses led to very different places; to valleys branching off in very different directions. They all led upwards to the great snow deserts of the Jostedal and Nord fiord, or to the dreary Sogne fjeld, and downwards again to rocky solitudes filled with the ruins that the recently receded glaciers had left behind.

He ascended one of the peaks in hopes of making a survey that would aid him; but found that the apparent summit was surmounted by another far above and far away, and that probably by another, and, perhaps, another still; as is often found to be the case in making such ascents. And he saw little more than peaks of rocks and plains of snow, and a portion of the *fond*, or motherland of glaciers, the vast table-land of snow and ice from which the numerous glaciers of this region descend. He descended and made for the pass that seemed the most likely to be the correct one. Of what he saw he writes:—

'On reaching the summit a singular scene presented itself. At the foot of a vast amphitheatre of snowy mountain peaks is a gloomy basin of rock, filled with the waters of a half-frozen lake. The water comes directly from the snow above, and is of a peculiar blue white, semi-opaque, London-milk colour, common to such snow water. This lake is called the Stiggevand, which, I believe, may be translated " Stygian Pool;" and a better name could scarcely be invented, for its gloom and desolate aspect would satisfy the imagination of the most dyspeptic and bilious of poets.

'The hollows, or basins, which occupy a higher level than the lake, are filled with snow and with ice formed by the melting and re-freezing of the snow. Thus filled up, they form great plains, having a surface of virgin snow, without a footmark, or a scratch, or spot visible. These apparent plains are, however, not quite level, but slope towards the rocky precipice rising above the lake. The ice sea, pressed forward by the mass above, flows over these walls in great bending sheets that reach a short way down, and then break off and drop in masses into the lake, their broken edges forming a blue cornice fringed with icicles. If these walls of the lake shore had sufficient slope to hold the icy cascade without breaking, glaciers would be formed; or if the supply of breaking masses were sufficiently great to overpower the thawing below, the basin of the lake would be filled up and become continuous with the great ice and snow-fields above, and might extend onwards to the spot on which I was standing, or even overflow this, and push down the valley up which

I had come. That this was formerly the case is shown by abundant evidences at every step of the day's walk, and the latter part of yesterday's.

'The soft, though sharp outline of the virgin snow, standing against the blue sky, just where it pours over the precipice, is very beautiful. There are no birds up here, no roaring torrent, no rustling of trees, no buzzing of insects; not even the ripple of a thin stream, as heard on the Swiss glaciers, but a silence that is almost absolute, and adds vastly to the effect of such a scene.

'The snow-plains, which are here seen bending over in cascades above the lake, are the northern terminations of the great table-land of snow forming the *fond*, or "Snee-fond," of the Jostedals Bræen, a great untrodden desert of perpetual snow and ice, extending for about fifty miles to the south-west, with a varying width, and covering altogether a space of about 400 square miles. Every valley of favourable configuration that branches from this great reservoir of ice is filled with a glacier or ice torrent, replacing the water torrents of the valleys that descend from the Dovre and other fjelds, that are not snow-covered.

'I now descended over similar ground to that on the opposite side of the pass. . . . I walked on over a wide field of glacier moraine, leading at last to the outlet of the Stigevand; a torrent of respectable dimensions which, fed by a succession of glaciers, grows to a river,* as it flows down the Jostedal. At the point where the stony fjeld narrows and descends to form the head of this valley the torrent makes a succession of falls over walls of piled-up boulders.'

---

* The Stor-elv, or Large River.

CHAPTER XVII.

SAETER LIFE.

OF most, if not of every one of the existing glaciers in Norway, and they are numerous, Principal Forbes has given an account with details, such as might be desiderated by the special student of their phenomena, but which are not so desiderated by the more general student of forest science; and sketches, not less graphic, of the phenomena presented by one and another of those connected with this *Sneefond* of the Jostedal Bræen, is given by Williams—amongst others of the Lodal, one of the largest, if not the largest, glacier in Norway, and of the Nygaard glacier; but enough has been adduced to give some idea of these feeders of the rivers.

Of the waterfalls details have been given sufficient to suggest some of the varied appearances presented by them; and of the lake scenery there have been given incidental notices, which leave them not altogether unknown. Bayard Taylor, in an account given by him of a trip to the Voring Foss, describes lake scenery and waterfalls in combinations which show each a character of scenery as a foil to the other. But the following description of the Nisservand lake, by the author of a volume entitled *Frost and Fire*, brings out as fully, and perhaps as effectively, the combination of lake and woodland scenery and may prove at once pleasing and more satisfactory. In the narrative referred to the author writes :—

'The *pasteur* (whose guests he and his fellow travellers had been) accompanied us to the beach, where we found a boat and two stout rowers in attendance. Pursuing our course up the Nisser-Vand, the western shore still continued to present the same bold and barren appearance;

but the prospect on the other side was enlivened by the frequent occurrence of hamlets and green pastures, occupying the gentle slopes of the hills. Every scrap of land, however small, that would afford footing to a goat, or space for a patch of potatoes, was taken advantage of. These little clearings, surrounded by the deep forest, and intermixed with crags and thickets had a most picturesque appearance. The marvel was how, with their utmost industry, the few roods of soil thus reclaimed could afford even a scanty subsistence to the population, which was evidently numerous. One might have wondered how access was obtained to these insulated settlements, shut in between precipitous cliffs above and the lake below, but that little piers and boat-houses under every cultivated nook indicated that its waters afforded the principal means of communication with each other and with the rest of the world. The winter, when it is one unbroken sheet of ice, must be the principal season for traffic and good neighbourhood.

'As we approached the head of the lake we were delighted with the series of dioramic views which the folds of the hills, stretching down in long slopes to the edge of the water, successively opened. In one place the bordering hills fell back, and left an amphitheatre of two or three miles in diameter, the undulating area of which gave to view the flowing lines of smooth and rounded masses of pines with which it was richly clothed, surmounted by bare cliffs behind; and over these, at some distance, rose a group of mountains of extremely fine contour, on the sides of which rested patches of snow at not a very considerable elevation. The lake terminates among a chain of low islets of graceful outline, some covered with young birch, feathered to the ground; others with a small clump of spruce firs, dropping their pendulous branches; some so small that a single tree only shot up its spiral form above the tiny patch of greensward that gave it footing.

'Threading our way through this bowery maze, we

landed at Vraadal, and, dismissing the boatmen, began the ascent from the shores of the lake, through open glades tufted with alder and birch. Looking backwards towards the bird's-eye view of the bay of islands now spread beneath, as on a map, with all the outline of bay, and inlet, and grassy promontory, and its network of intersecting channels gleaming like silvery threads, and opening out into the last broad reach of the noble lake, was a scene of indescribable beauty. Before us, at a great distance to the north-east, stretched away the dark chain of mountains which bounds the valley of the Maan, towards which our course was tending. In about two hours, descending through a spruce grove of particular luxuriance and very lofty growth, we caught glimpses among the trees in the windings of the road of a fine sheet of water below, and of an enormous mountain mass, which rising directly from its edge, towered to a height of 4000 feet above the fiord.'

In the higher-lying plateaux, whither we have been led, there is carried on an interesting phase of rural industry, which has been alluded to in some of the narratives which have come under consideration in preceding statements in which mention has been made of *Saeters*, or cheese-making establishments on the mountain pastures. These are generally at a considerable distance from the farmhouse, and thither are sent the milch cattle under the charge of young women, there to spend the summer months, where, for the accommodation of the dairy maids, and the prosecution of their dairy operations, there are huts bearing the designation I have cited.

There, is seen a phase of rural life extensively practised in Norway, and not unknown in Sweden, so distinct in its usages from corresponding practices followed in dairy establishments in other lands that it may be cited as one of the national peculiarities.

These saeters are one-storeyed huts, each containing two rooms. The outer apartment, which is fitted up with a hearth, a table, and a coarse bed, is the living room; the inner one is the dairy, containing the cheeses and the implements used in the manufacture of them.

Often several, belonging to different proprietors, form a group in a mountain hollow; sometimes you see them by the shore of a tarn. Within a few miles from Christiania is Frogner Saeter, the property of Herr T. J. Heftye, Consul of the United States, and President of the *Turistforening* or Tourist Society, an association formed to foster a taste in the country for mountain exploration, to facilitate which they have, at considerable expense, had mountain roads made and improved, and huts erected in remote localities in which members of the society may find shelter and rest on their extensive tours.

Frogner Saeter is situated some 1,700 feet above the level of the sea. The approaches to it are through a large and dark forest by a road made at great expense by the owner. From the saeter a magnificent view may be had. Another mountain lodge belonging to the same proprietor is Sarabraoten, situated in a wild region in a romantic spot, overlooking a picturesque lake. His love of wild scenery has prompted him to build at both of these places houses like those constructed in the olden time; but to see saeter life in its reality the traveller must go much further afield. In most of the published journals of tourists in Norway may be found accounts of visits made by them to saeters. An interesting chapter in Du Chaillu's volume, entitled *The Land of the Midnight Sun,* is devoted to an account of a visit which he made for some days to a saeter in the Valdal.

'In the midst of the mountains,' he writes, 'Far away from the farms, by the shores of lonely lakes and rivers, or on the slopes of ridges beyond the limits of the growth of grain, are the saeters. These are mountain houses or huts built of logs or of rough stones, where during the summer months the people of a farm come to pasture their cattle, for in the midst of this great wilderness of rocks there are many spots covered with aromatic grass, which gives a rich flavour to the milk of these places. Many of these saeters are very difficult of access; high mountain ranges and snow patches have to be crossed, and rivers forded by man and beast. Solitary indeed is the life in these mountains, for only once or twice during the summer does the farmer go up there to see how those who have been left are getting on, to hear about the herds, and if the season has been good. . . . On these visits they bring provisions, and take back the produce of the dairy. The saeter life is also a hard one; the pastures are far away from the huts, and during the whole day the maidens have to follow the herd, rain or shine, and return in the evening, cold, hungry, and often wet.

'In some mountains pastures are very abundant, and saeters are numerous; in others they are few and far apart. Almost every farmer possesses one, but some who have more mountain land than they require rent part of this to those less fortunate. . . . The people start for the *saeter* in many districts toward the middle of June, the time varying somewhat, but generally not after midsummer—St. John's Day—according to the distance and the mountain heights that are to be crossed. They return between the middle and end of September, and if high mountains are to be passed, about the first week in September.

'The young maidens, the pride of their family or of a neighbourhood, will remain in the mountains all alone feeling as safe as in their father's home; they have no fear of being molested, for they trust to the honour and manhood of the Bond—agriculturist—blood. Very few things in Norway have impressed me more than this simple faith.

'The young lover comes once or twice to cheer the hours of his sweetheart, but only for a day. If engaged to him he is the more welcome, for in the autumn, after the labours of the harvest are over, the wedding will probably take place.'

In connection with this I may cite the following information given by him in another connection:—' A betrothal in Scandinavia is celebrated in a festive manner. In the country districts the engaged couple often go before the clergyman, who, in presence of the respective families, says:—" Before God, the All-knowing, and in presence of these witnesses, I ask thee if thou wilt have, him or her, for thy betrothed?" After an affirmative answer from both, rings are exchanged as a pledge; these are worn on the ring finger of the left hand. The custom of going before the parson is dying out. In cities, or among the educated classes, after a gentleman and a lady have become engaged both their names are written on a single visiting card and sent to all their acquaintances, this being a notice of the betrothal; it is also published in the newspapers. The lady, after her engagement is announced, is allowed to go with her affianced, and they are often seen together without their families at balls or places of amusement. Nothing but a plain gold ring is given even among the most wealthy. The wedding token is of the same character. When a woman has a family she wears three rings as a mark of distinction, of which many feel very proud, though this last fashion is going somewhat into disuse.

'A few days before the departure for the saeter a great stir takes place on the farm; milk-pails, churns, and wooden vessels, the great iron pot, the mould for the cheese, two or three plates, and a cup or two, a frying pan, and above all the coffee kettle, are made ready for packing. Salt for the cattle, flour to be mixed with skim-milk for the calves, bread, a piece of bacon for Sunday, coffee and sugar, a covering for the beds, must not be forgotten. The girls take their Sunday clothes, and prayer-books, and old garments for every-day use; a good stock of spun wool to make stockings, mittens, or gloves, in their leisure hours, and pieces of cloth upon which they can embroider. The old horse which carries the load is often let loose to pasture in the mountains for several weeks, for the ploughing is over, and the grass or hay left from the year before is carefully saved.

'On the morning of departure the cows, sheep, goats, and a pig or two are watched by the children to keep them from straying far away. If the farm is small, and the people poor, all the family go to the saetar till harvesting takes place, The mother is often seen carrying on her back the last baby. Before starting the mother prepares an extra good meal for the farm-hands, or that part of the family who are to accompany the saeter girls—the daughter or girls hired for the season. Those who take the lead often carry long horns by whose shrill tones the animals are called to follow, salt being given to them, now and then, to coax them on, and the children keep them in line.'

Of his visit to the saeters he tells:—' In the beginning of July I left the old city of Stavanger. . . . The sail on the fiord was very interesting. After a trip of twelve hours we came to the end of Sands fiord, a branch of the Stavanger; here I landed with my guide, Samson Fiskekjoen, who had been recommended to me as trusty, and well-acquainted with the mountains. . . . After a drive of two hours through the picturesque valley of Suledal, along the clear river, we reached his farm, where we found his father, then eighty years old, splitting wood with a strength which augured well for a life of twenty years at least. The old couple received me with great kindness.

'A number of farms were scattered about, and in sight was the church; a short walk brought me to the parsonage. . . . During my absence a complete metamorphosis was effected in the farm-house, and everything was tidy and clean; bread, butter, cheese, and sour milk were on the table, and the good people excused themselves for having no sweet milk, as the cows were far away in the mountains. I slept with my door wide open, for the night was very warm; I do not think they slept at all, as coffee was ready for me at four in the morning. They pressed me to eat, as the journey before me was a long one.

'I left with two boatmen. We had not sailed far before we came abreast of a comfortable white-painted house, the pleasant home of a Stoi thingsmand, where we went ashore. The host was not at home, but his amiable wife, who had heard of my coming this way, had been expecting me, and seemed quite disappointed when she heard I had spent the night at the farm of Samson. Though I assured her that I had breakfasted she insisted that I should partake of another.

'The Sulsdal valley, near the lower extremity of the lake, is exceedingly interesting to the antiquarian, on account of the numerous tumuli, or tombs of heathen times, some of which are hollow, or circular in shape, and surrounded by stones, while others are square. As we ascended the lake we could see the paths leading to the saeter, and patches of snow on the mountains. After a pull of fourteen miles we landed at Naes on the right shore, near the upper extremity of the lake, from which there is a horsepath leading to the numerous saeters met between the Suledal and the Poeldal lakes. . . .

'The path, after leaving Roeldedal, ascended gradually along the Valdal river, in view on the left bank of the white column of the Risp-Foss; descending again and crossing the stream on a bridge, we saw on the opposite shore the bridle-path going to Lake Staa and Upper Thelemarken.

'On the right bank of the Valdal are seen many saeters and paths branching in every direction. The river flows for some distance through a flat country, dotted with fine pastures and small farms. Another stream throws itself into the Valdal, and forms a magnificent cascade of 1000 feet, below which the current was so strong that even the horse could hardly keep his footing while fording it. Twelve miles from Foeldal we came in sight of the Valdal lake, the mountains sloping gently to the shore, near which were several saeters. Herds of cattle which had come from the mountains to be milked grazed on the green banks, and on our left high up was the Barken saeter; while at the head of the lake the smoke curled upwards from the Valdal saeter, and we heard the loud cries of the girls calling the cattle that wended slowly on their way, browsing as they went. On the right bank of the lake a magnificent cataract fell from a great height.

'We followed the shore till we came to the upper extremity of the lake. The people were watching us, wondering who we could be, for they expected no one from their home.

'On our arrival they bade us enter the house, which was as comfortable as that of a farm, and the usual salutations took place; the milk was passed around in the large flat pail in which it is kept for the cream to rise; taking the customary sip we handed it back with thanks, and the usual pressing invitations to drink more were responded to by drinking as much as we could with Many thanks— *Mange tak.*'

There, as elsewhere in Norway and in Sweden, he was made the more welcome when they learned he was from America where a member of the family was settled. 'The father had come the day before to carry back the butter and cheese which was made to the gaard or farm, which was at a great distance on the Soer fiord, one of the branches of the Hardanger. He was the father of a large family of grown-up children, . . . a type of the Norseman—North-man—hospitable but undemonstrative, with a tall and spare figure, and a kind face. Three of the daughters were at the sacter for the summer, all of them pictures of health, and blondes of the type of the descendants of the fair-haired Vikings. Syvnor, the eldest, rather short in stature, was nineteen years old; Anne was seventeen, tall, muscular, with piercing blue eyes, and fully able to take care of herself: she would have made a good model for a valkyrie; Martha was sixteen, with golden hair, soft blue eyes, and delicate complexion. All three were celebrated on the Hardanger for their beauty, and young farmers, without number, were trying to win their hearts.' He praises the maidens, and the invigorating climate prevailing at such places 4000 feet above the level of the sea; and he goes on to tell:—

'The mountain life is an active one, and the girls are busy from sunrise to twilight. The pastures belonging to this saeter were extensive in the neighbouring mountains, and sufficed for fifty-two milch cows, with eight others, and four horses. The cattle belonged to three different farms, including that of Nels. The farthersome coming from Soer fiord, fifty miles distant; two of his daughters had charge of those not belonging to him, for which they were paid. The milk of each herd was put in the vessels belonging to the place from which the cows came, and the butter and cheese were set apart in like manner. The people are so honest that no farmer fears that the girls will favour one at the expense of the other, or put any of the butter or cheese in vessels belonging to any but the rightful owners.

'A large enclosure, surrounded by a stone wall, contained a fine meadow, the grass of which was carefully cut and dried, to be taken away by sleighs in the winter. There were upwards of 250 milch cows at the Valdal saeters, besides large numbers of heifers, calves, and horses. The calves were kept at home; every morning and evening they were fed on a mixture of churned milk and flour, with salt; or if no milk was to be had, on hot water, in which juniper shrubs had been kept for a while.

'At four o'clock in the morning we were awakened by the ringing of the bells, which some of the cows wore around their necks; they had come by themselves from the mountains to be milked, and this was the signal for the girls to rise. This they did at once, and were soon on duty. . . . Each buckling on her waist a belt from which hung a horn filled with salt; this is given to the cows as well as to the horses and sheep, generally in the morning and evening, when they go to or from the mountains.

'After the milking the girls drove the cows up another path in the mountains to new pastures, from which they would come and go by themselves after knowing the way. On their return the maidens went into the milk-room, the door of which was always carefully closed, skimmed off the cream which had been formed on the milk of previous days, and putting it in the churn, they began to make the butter. Others took the empty vessels to the river, and rubbed them inside and outside with fine sand from the shore, and afterwards with juniper branches, finishing by a thorough rinsing in the stream. The pails are generally made of white pine, and are clean and spotless. Cheese day also proves a busy time, and its work is done in the same thorough manner. The room where the milk is kept was marvellously neat; about 150 pails filled with it were on the shelves, each being about twenty inches in diameter and five inches deep, made of white pine, with wooden hoops; the milking-pails stood on the floor ready to be used. Several barrels for the churned milk and butter-milk, and vessels for the butter, were also arranged in order.

'On Sunday, after their morning milking, every one commenced his or her toilet, as if getting ready to go to church, putting on clean linen, and all their holiday clothes and shoes. The girls and their mother wore dresses of thick dark-bluish woollen material, homespun, with corsages of the same colour. The bottom of the skirt was ornamented with a wide green band all around. The corsage was open, and showed a handkerchief embroidered with gold. Each girl wore a close-fitting little cap, which seemed to be made only to hide the ends of her thick luxuriant hair. No work was done, except what was absolutely necessary; some of the family read the Bible and sung a few hymns of praise. After dinner visiting took place from saeter to saeter, and the afternoon was spent in the social fashion customary to the country.'

Du Chaillu went on such a visit with one of the maidens. 'Early on Monday morning,' says he, 'Everybody was up; the horses were ready for the return of Nels to the farm; the pack-saddles were put on over two thicknesses of woollen blanket; the butter, cheese, and milk for the working people on the farm were not forgotten; the father, in a quiet way, without kissing, said good-bye to all his family, and soon was lost to sight in the windings of the path beyond the lake.' Du Chaillu also resumed his journey to visit other saeters.

'From Lake Valdal,' says he 'The path northward, over the mountains, is wild and dreary, even in the beginning large patches of snow having to be crossed.

'After leaving the lake we ascended over a rugged country above the birch region where juniper and Arctic berries were abundant. An hour's walk brought us to the shores of the lakelet Visadal Vand, not far from which was an isolated poor-looking saeter built of loose stones. The inside was far from clean; on one side were the beds, placed on the rough slab-floor; on the other the fire-place; in a corner lay a heap of juniper bushes, five or six pails, a copper kettle for making cheese and boiling milk, a coffee-pot, and a churn. The occupant of the saeter and his wife welcomed me; the man was apparently more than eighty years of age, but hale and hearty; he had travelled about eighty miles to spend the summer here, and well exemplified the hardiness of these mountaineers. This saeter had 120 dry cows, belonging to many farmers, who had sent them here to pasture. A hired woman and three men had the charge of them, having also five milch cows for their special use beside.' He and the guide continued their journey over bare rocks and patches of snow, sometimes the horse taking one way up the steep ascent, and they another, passing many cascades and waterfalls. Others have told me the same tale. He writes, . . . 'We were still ascending, and our pass was more than 4000 feet above the level of the sea. The fields of snow, which were deep and soft, increased in size, and we had to cross one, horse and all, almost one and a-half miles long: now and then we saw tracks of wild reindeer. Suddenly we found a track of red snow in the midst of the white, the first I had ever seen. I imagined a reindeer had been killed there, and that the snow had been stained by its blood.' 'This is gammel snow—old snow,' said my guide. As we advanced these rose-coloured patches became more numerous. Some of them being fifteen feet long, the effect was very striking. This red snow is always found in the large melting patches, and its colour is due chiefly to the presence of minute vegetable organisms, enclosing an oily-like red liquid, this alga is known as *Haematococcus—Protococcus nivalis.* We then passed on the border of Vasdalseggen, where the mountains, largely covered with snow, range in the direction of north north-west. After we had traversed this plateau for about three hours it slopped downwards to the east, and a toilsome tramp through wet snow brought Lake Bjoerne into view. On its shores I saw cattle grazing, and not far off the smoke curling from a solitary *pige saeter*—girl saeter, in this mountain home of the wild reindeer.

'Every year, towards the latter part of June, from the Hardanger fiord, or from Roeldal, a farmer, accompanied by two girls, with a drove of milch cows crosses these mountains. During the summer the girls are left to take care of the cattle, and to attend to the diary.

'It was late in the day when we arrived at this lonely place; the girls came out to see who the strangers were, suddenly disappearing at our approach, to put on their best clothes to receive us. They wore the costume of the girls at Roeldal, and their caps were set very coquettishly on their heads. One had red stockings, the other blue.'

M. Du Chaillu goes on to describe the life and occupation of the girls, which were similar to those of the first saeter visited by him. I have been pleased with what I have heard of the chivalrous respect with which the saeter girls are treated by young men, their acquaintances, occasionally visiting them; and with what has been told me of young women completely non-plussing young men who had presumed to think they might be trifled with with impunity. One scene flashes itself upon my memory: a brave maiden raising into threatening position the handle of her broom, and, with contempt in look and tone, speaking to the cowed bully, much as God spake to Sennacherib,

'The virgin, the daughter of Zion, hath despised thee, and laughed thee to scorn; the daughter of Jerusalem hath shaken her head at thee,' while the would-be bravado looked like a whipped cur.

'In Norway,' it is said, 'butter and cheese are much used as food. There are three peculiar kinds of cheese:—1. The *Mysost* is made from the whey remaining from the common cheese, boiled till the water is evaporated; then it is shaped into square cakes, weighing from two to five pounds; the colour is dark brown. It must stand at least a day before it is fit to be eaten. It is only made at the saeters where wood is plentiful, for it requires a great deal of fuel. It is eaten in thin slices, and with bread and butter; women and children are especially fond of it. The best is from goats' milk. It should hardly be called a cheese, as it consists chiefly of sugar and milk. 2. *Camel-ost*, made from sour skimmed milk, is a fermented round cheese which is kept for months in the cellar. 3. *Pult-ost* is also a fermented cheese mixed with carraway seeds, not formed into cakes, but preserved in wooden tubs.'

The *Mys-ost* has a sweetish flavour. It is the colour of Windsor soap, and nearly the shape of a brick. The *Camel-ost*, or old cheese, when good, reminds one of stilton in the very last stage of decay.

CHAPTER XVIII.

VALLEYS.

DESCENDING from the high-lying plateaux with their snow-fields and glaciers, and passing saeters scattered among the hills, we find we are brought by a succession of valleys towards the coast. As we journey adown these valleys it may occur to us that the origin of them we have seen in the apparent level of the fjelds from which we have come.

Of the plateau of Southern Norway, Dr Broch writes: 'From this plateau issue all the great water-courses and rivers of Southern Norway. They take their rise in part from marshes, in part from the névés, in part from deep basins without any apparent affluents, in part from lofty eminencies where the sources are unseen, but where are constantly being condensed the currents of humid air coming from the west. At their birth the rivers wind about on the plateau in innumerable small tortuous furrows, which go from pool to pool, from hollow to hollow, from lakelet to lakelet. Having attained mediocre development, they rush along in sinuosities from pond to pond, and at length from lake to lake. These reservoirs follow in succession, like strings of pearls, the smaller they are the closer are they together. The more the land is cut up, the deeper also are the waters, and in general the more restricted in size. And it is no small portion of the area of the plateau which is covered with stagnant water which, as do also the running waters, abound in fishes and frogs.'

These water-courses are incipient valleys—they are valleys in the plateaux—valleys which may yet become in the course of ages wider and deeper, with well-defined slopes, and already they are the upper portions of the valleys which lie far beneath, into which they lead; and the *thalweg*, or valley-course, may be traced by the continuous flow of the water by the rivers, from these streamlets on the plateaux to the mouth of these, by which the water of a million rills, or a million million of them, flows into the sea.

Treating lakes as we have done, as expansions of rivulets or rivers filling some hollow or valley, we come naturally to consider valleys as river-courses and lake basins; and such they are—in some few cases having supplied water leadings and reservoirs; in most, having been fashioned, if not also formed, by the flow of water or of ice—formed, it may be, by the latter, and fashioned by the subsequent water-flow. In the descriptions given both of rivers and of lakes we have found the setting of the picture to be valley scenery; but the valleys may be described apart.

Williams, a graphic writer, of whose word paintings I am glad to avail myself, supplies an illustration. In continuation of a narrative which he gives of a journey which he made across the Dovrefjeld and into regions beyond, he writes: 'The road beyond the station [Bjerkager] commands fine views of the valley, a deep ravine, thickly wooded with fir trees, and the river dotted with pine-covered islands. There are many indications of glacier action hereabouts, similar to those in the valley of the Driva, but more extensive and decided.

'The rich verdure of the Guldbrandsdal prevails over the greater part of the country through which I have walked to-day, and the fields are carpeted with sweet flowers as were those of yesterday. I little expected to find this element of beauty so generally prevalent in the far North.' Passing Soknaes station he writes: 'The road now enters the Guldalen, or valley of the Gula, the view down which is very beautiful. It is a rich cultivated valley, the river winding through a fine wooded plain, and round about green knolls and mounds, that have a very complicated appearance even from above. On descending the valley, and walking a few miles down it, the structure upon which the peculiar appearance depends becomes evident.

'There are two very distinct kinds of valleys commonly met with in mountainous countries : one, the long narrow ravine, a mere stone trough, formed by the rocky slopes of the mountain sides meeting each other at an angle; this angle being more or less choked with fragments of fallen rock, among which a torrent roars. These valleys vary very considerably in their features according to their elevation, the steepness of their sides, and the character of the rock composing them. Some are deep gorges, with barren and almost perpendicular walls; others have a more gradual incline, and their sides are covered with woods, or cultivated ledges and slopes. The other is the open, basin-shaped valley. This, like all valleys of any considerable extent, gives path to a river or small stream; but if the wide basin-shaped valley be deepest in the middle, as is usually the case, the river fills the hollow, and forms a lake, spreading itself out in calm repose after its fitful journey among the rocks above. Thus, the lake of Geneva is the sleeping Rhine. That of Constance is the Rhine reposing in like manner. The Mediterranean is a larger valley filled with waters, where many rivers sleep. And the ocean is the main valley of the world, the final resting-place of all the rivers.

'There is another modification of this open basin-shaped valley, where a lake of earth, generally fertile soil, takes the place of the outspread river. This is easily accounted for. The toiling river brings a burden with it, which it lays down at its resting-place. So long as it continues in rapid motion, stirred and eddied by the resisting rock, it is turbid and milky with the suspended particles it has abraded from the mountain sides, but when it becomes quiescent, these sink to the bottom, the largest first, and so on; and the river issues from the other end clear and refreshed, ready to resume its levelling labours lower down.

'By such a process of deposition are these wide valleys gradually filled up, and then the river flows gently in a long winding course through the rich territory of its own depositing, like an old man calmly enjoying the fruits of his early toil, and contemplating the good deeds of his youth; for the youthful river, in the brawling early days of its mountain life, is doing good service to the world in thus converting desolation into fertility. Nearly all the fertile plains of the earth have been created thus by the industry of rivers.

'Besides these there is seen yet another kind of valley, partaking of the characters of both of the above: a long trough-like valley formed by the mountain sides, but which widens as it proceeds downwards, and branches into the great valley of the sea. The waters of the sea fill its lower part, and an estuary, firth, or fiord is formed. These in like manner are continually being filled up by the rivers which come to rest in the waters of the sea, and deposit their burdens there. Thus has Holland, the master-piece and last labour of the Rhine, been formed.

'The Gula, into whose valley I now descend, presents some illustrations of these river agencies, and a problem to boot. The mounds and knolls that appeared so complex from above are seen from below to be formed by the river cutting its way through the alluvium it has deposited. This may have been effected in two ways; the deposit may have been made in a lake filling a basin-shaped valley, and the river may have cut down and lowered the channel of its outlet considerably beyond its original depth, and thus have not only drained the waters of the lake, but have given sufficient inclination and velocity to the river to enable it to carry with it much of the soft earthy matter over which it was flowing; or it may be that this was an estuary valley, an ancient fiord, up which the sea stretched an arm, the alluvium being deposited by the river when it entered the

O

sea; and after this was done, the whole valley, mountains, river, and its deposits were all uplifted by the fiery forces within the earth, which battle against the working of the waters outside, raising new mountains while the waters wear the old ones down. Such an uplifting would lengthen the journey of the river, as the sea rolled back from the uplifted land. In its new course the river would cut through the soft plain it had formerly deposited on the bottom of the ancient fiord, and continue cutting down till it reached nearly the level of the sea; and thus the depth of the cutting would measure the amount of uplifting.

'Throughout nearly the whole of to-day's walk—about twenty-five miles—terraces formed of alluvium were visible. In some parts the river flows at the foot of a steep bank of even slope, above three hundred feet high, the top of which is a cultivated or wooded level; at other parts there are several step-like terraces, parallel roads, as they are called in Scotland. Near to Medhaus station I counted five of these, one above the other, and perfectly parallel. From the course of the river, and configuration of the valley, I suspect that these terraces have been formed in an estuary which has been rendered high and dry by the uplifting of the land. If so all the neighbouring valleys that carry considerable rivers into the sea should present similar phenomena, more or less distinctly marked.' That this is the case with some of them, I knew from reading the accounts of other travellers. It is the general opinion of geologists that the whole of Scandinavia has been uplifted at a geologically recent period.

CHAPTER XIX.

FOREST EXPLOITATION, TRANSPORT OF TIMBER, AND EXPORT TIMBER TRADE.

FROM a friend who has travelled more extensively in Norway than I have done, I have learned—and the information is in accordance with what may be seen from numerous forest maps which have been issued by the Government—that most of the forests are found along river-courses. They extend from half a mile to three or four miles from the banks of the river, and up the precipitous hill-sides beyond. Sometimes the continuity is broken abruptly on the river-bed by perpendicular cliffs; but the forest extends on the table-land above, like a dislocated geological stratum, or the further side of a dyke or fault, and this gives, as has been stated before, a character to the exploitation of the forests, in connection with the bringing out of the timber.

Many of the forests are private property; others belong to commercial proprietors. In both classes of forests the right to fell timber is generally let to contractors possessed of large capital, by whom arrangements for felling wood upon an extensive scale are made.

Previous to the introduction, of late years, of an improved forest economy, the system of exploitation or working usually adopted was one intermediate between that known in France as *Jardinage*, felling only such trees as were desired, and that known as *A tire et aire*, in which the forest is divided into as many sections as periods required for the reproduction of the crops, and these are cleared in succession, but only one in each period: the *coupés*, or fellings in different periods in these Norwegian forests not being regulated in extent by precise measurement, but being determined by the convenience of the contractor; and only trees suitable for his purpose being felled. These are generally trees, on an average, a little under two feet in diameter; and all such are felled, leaving after them but a poor and scraggy crop of growing trees to replace, in course of time, if they can, what has been removed.

In Norway there is no lack of means for transporting the felled timber by water to the coast. In many places the felled trees, stripped only of their larger boughs, are tumbled into a mountain stream, to be by it borne to the nearest river or lake; in others, they are shot along artificially constructed slides, leading to some lake or river. These slides are in structure intermediate between the *chemins à trainaux* and the *lançoirs* or *glissoires artificiels*, used in France. They are about 5 feet wide. Sleepers are laid across the line at about equal distances apart, and upon these are laid, lengthwise, trunks of young trees about 5 or 6 inches apart, and often so arranged that those at the sides are somewhat higher than those in the middle so as to form a groove of sufficient depth to keep the shot timber in the slide. In some cases these slides run directly down the declivity to the river or lake to which they are destined to convey the timber. In other cases, they run across the side of the hill in a slanting direction. In some places earth is removed to allow of the desired angle of inclination being secured. More frequently this is attained by the slide being supported at places by piles of earth or beams. When necessary, they are carried on supports across small valleys, or watercourses, separating the forest on the one side of a mountain from the forest on the side of another; and occasionally there may be seen their straight course altered only by an angle more or less abrupt. At such places there is generally raised at the outer angle of the slide a bank against which the trees may strike in their descent and then recoil into the new direction; these by the new direction thus given to their course, go on, sometimes head foremost, and sometimes making first a complete somersault or revolution. In general also a workman, or it may be two, or even three, are stationed at these points with long poles to aid at the time the movement of any trees which might otherwise be in danger of sticking fast and blocking the way. The slides in general lead to pools of considerable depth in a lake or river. Into these the trees ofttimes descend more than their entire length, starting up again vertically before setting off anew on their course.

Much *débris* is found all about such spots; but it is comparatively seldom that logs are seriously damaged. The quantity of splinters may be, to a great extent, composed of the lesser branches left on the tree when placed in the slide.

Brands or marks may be seen upon some logs. These are, I presume, the marks of the woodcutter or the contractor, made to enable each one to claim his own property should logs belonging to different proprietors get mixed together.

The logs are carried down by the river, and if the river falls into a lake, they are—at the embouchure of the river—collected and formed into a raft; and such rafts are sometimes towed by a little steamer across the lake to its outlet. If the stream flowing thence be smooth, they may be floated further as a raft; if it proceed over waterfalls in its course, the logs are unchained and allowed to float down apart, to be reformed into a raft below these, if circumstances allow of this. Notwithstanding the care which may be taken, many logs are stranded on the banks of the lakes and rivers. The logs are cut into size and shape for the foreign market by saw-mills near the coast, which are driven by water-power.

Saw-mills of the simplest structure, consisting only of a water-wheel and a circular saw, fixed apparently on the same axle, are common appendages to farmhouses in the country. They are employed in cutting up the firewood required for the family and their retainers.

By Forrester it is mentioned that in some cases two years have been occupied in the transport to the sea of the timber cut in the upper mountains. From this some idea may be formed of the difficulties which have to be overcome, and which are overcome, by the indomitable industry of the people.

I have stated what I saw and learned of timber floatage Torristal, or Otter-elv [ante. p. 4], and what I saw of the floatage of timber on the Glommen [ante. p. 10].

There, as elsewhere, logs are transported from the spot where they are felled to the banks of the nearest stream, and marked with the initials of the owner. On the melting of the ice they are pushed into the current, and the contributions of many affluents find their way to the river, which may at the time be covered with the floating masses, which become more or less compactly interlaced, till some projecting rock in the bank or the river bed arresting some, others are impeded and stopped in their course, and ultimately many thousands, it may be, are stopped, and piled up in a confused heap. It is perilous work to break up the piled mass, and set the logs afloat upon the stream again. In doing so, ' the men employed go about balancing themselves on detached logs in the middle of the stream, pushing on each log by means of a boat-hook, till at last the mass of logs hanging together begins to be disturbed and shake, and then comes the struggle for the men to regain the shore. The skill which the men display in disentangling the logs, the agility with which they run about and maintain their balance on the floating logs, as well as on those which are fixed, the intelligence which they apply to the separation and setting afloat again of all those interlaced logs, and, in fine, the courage with which they face all these perils, are all of them worthy of admiration.' The statement is cited from the report by Dr Broch.

The author of the work entitled *Frost and Fire*, to which I am indebted for the sketch of logs performing the

Halling dance below the waterfall on the Torristal river, some distance above Christiansand, tells that after the logs have been launched 'many get waterlogged and sink; and these may be seen strewed in hundreds upon the bottom, far down in clear green lakes,' and he goes on to say:—

'Many get stranded on the mountain gorges, and span the torrent like bridges; others get planted like masts amongst the boulders; others sail into quiet bays, and rest upon soft mud.

'But in spring, when the floods are up, another class of woodmen follow the logs and drive on the lingerers. They launch the bridges, and masts, and stranded rafts, help them through the lakes, and push them into the stream; and so from every twig on the branching river floats gather as the river gathers on its way to the sea.

'Sometimes great piles of timber get stranded, jammed, and entangled upon a shallow, near the head of a narrow rapid; and then it is no easy or safe employment to start them. Men armed with axes, levers, and long slender boat-hooks, start down in crazy boats, and clamber over slippery stones and rocks to the float, where they wade and crawl about amongst the trees, to the danger of life and limb. They work with might and main at the base of the stack, hacking, dragging, and pushing, till the whole mound gives way, and rolls and slides rumbling and crashing into the torrent, where it scatters and rushes onwards.

'It is a sight worth seeing. The brown shoal of trees rush like living things into the white water, and charge full tilt, end on, straight at the first curve in the bank. There is a hard bump and a vehement jostle; for there are no crews to paddle and steer these floats. The dashing sound of raging water is varied by the deep musical notes of the battle between wood and stone. Water pushes wood, tree urges tree, till logs turn over and whirl round, and rise up out of the water, and sometimes even snap and splinter like dry reeds.

'The rock is broken, and crushed, and dinted at the water-line by a whole fleet of battering-rams, and the square ends of the logs are rounded; so both combatants retain marks of the strife.'

Like pictures might be cited from *Rambles in Norway,* by Thomas Forester, Esq., and others.

In the *Revue des Eaux et Forêts* is given an extract from a report of M. de Reujoux, Consular Agent of France at Christiania, On the exportation of wood from Norway, which had appeared in the *Bulletin Consulaire Francais,* of which the following is a translation:—'The exportation of wood from Norway in the year 1880 amounted to 894,816 tons. England occupies the first place, with an increase over 1879 of 156,498 tons— viz., 33,345 tons of planed wood, 37,564 tons of sawn wood, 85,589 tons of spars and mine props of two dimensions. Then comes France, with an augmentation over 1879 of 26,567 tons—viz., planed wood, 4,355 tons; sawn wood, 17,967 tons; beams, 2,639 tons; staves, 1,608 tons. The exportation to Germany has considerably fallen off, owing to the rigorous enforcement of protective duties since the 1st October 1879, leading to a decrease of 23,599 tons in 1880, chiefly in planed wood.

'Norway has found a new outlet for its timber produce in Holland, in consequence of the abolition of impost, which impost formerly rendered exports to this country almost impossible. During the year 1880 about two-thirds of the wood sent to Holland merely passed through it in transit to Germany. Consequently, of the 38,942 tons sent to Holland, only 19,862 tons could be assigned to the trade in Holland.

'With regard to Australia, it is impossible to supply precise numbers, seeing that England sends thither much of what she herself imports from Norway. The direct exportation to Australia (11,575 tons) is considerably above that of 1870, which only amounted to 7,787 tons. The last three years (1878, 1879, and 1880) give a mean of 6,382 tons.

TRANSPORT AND EXPORT TIMBER TRADE.

'With Africa since 1870 transactions have been continuously augmenting.

1870	. . . . .	823 tons
1879	. . . . .	3,806 ,,
1880	. . . . .	7,646 ,,

'Spain is of no importance to Norway in connection with timber exports, nor will be so long as the present system of taxation there is continued.

'With Belgium business has manifested some development in 1880, but the impost duty still shackles it considerably.

'It follows from a comparison of the totals of the quantities exported during the years 1879 and 1880 that the export of wood during 1880 has exceeded that of the preceding year by 187,546 tons which is divided in so far as the different kinds of wood are concerned as follows:—

KINDS OF WOOD.	1880. Tons.	1879. Tons.	Incr. 1880. Tons.	Decr. 1880. Tons.
Planed Wood, No. 1	193,654	164,682	28,972	—
Sawn Wood, No. 2	245,548	176,893	68,655	—
Spars, Mine Supports, Nos. 4-8, 10, 11	295,616	207,497	88,119	—
Beams, No. 3	88,519	102,054	—	13,535
Staves, Nos. 12, 13	30,061	26,148	9,913	—
Other Woods (Danish No. 9)	29,576	29,996	—	420
Laths and Firewood, No. 14	11,842	—	11,842	—
Totals,	894,816	707,270	201,501	13,955
Difference in excess, 1880,	...	...	...	187,546

With the exception of the years 1871, 1872, 1874, and 1876, Norway has never exported so much wood as in 1880. The exports of late years are otherwise shown by the following table:—

1880,	. . . .	894,816 tons.
1879,	. . . .	706,950 ,,

1878,	.	.	.	.	737,014 tons.
1877,	.	.	.	.	830,508 ,,
1876,	,	.	.	.	932,654 ,,
1875,	.	.	.	.	751,309 .,

'Since 1870 the exportation of planed wood shows a continuous augmentation. In 1880 it amounted to more than one-fifth of the whole export, showing an increase, which was progressive, from 96,445 tons in 1870 to 193,654 in 1880. Sawn wood, on the contrary, has fallen from 428,553 tons in 1870 to 245,548 tons in 1880. Spars and mine supports have increased from 100,552 tons in 1870 to 295,616 tons in 1880. Lastly, staves have increased from 19,631 tons in 1870 to 30,061 tons in 1880.

'The six principal Norwegian ports, arranged according to the importance of the wood exports, give for the years 1880 and 1879 the following results:—

	1880. Tons.	1879. Tons.
Fredrikstadt	132,383	107,244
Christiania	126,234	90,838
Drammen	117,270	87,783
Christiansand	53,739	41,994
Fredrikshald	53,436	34,583
Arendal	50,585	46,038
Totals	533,647	408,481

'It may be seen from this table that the town of Fredrikstadt occupies for the past, and it always will occupy, the first place, on account of its advantageous position at the embouchure of the Glommen. Indeed, the general report for the year 1880 shows that in the basin of the Glommen there have been marked by the traders 185,395 tylters and three pieces of trees. (A tylter contains twelve trees, felled and bound together.) These 185,395 dozen trees have been floated and delivered to the owners at their mills, with only a loss of 3 per 200. This is far beyond what they had expected, according to the prospects suggested by the spring and summer of the year preceding.

'The good result has been equally complete in the

greater part of the water-courses utilised for floatage, so that of the 680,000 dozen nearly which have been floated throughout the entire country, there has remained scarcely 8 per cent., or 54,000 dozens, of which 30,000 dozens belong to the basins of Krageroe, of Arendal, of Christiansand, and of Mendal, where there were marked about 125 dozens.

'The mean gross price of the wood floated in 1880 in the basin of the Glommen (calculated according to the results yielded by more than 100,000 dozens) was 25 crowns 26 ore, about 26s 6d sterling per dozen. Consequently there had been paid to proprietors of forests in the basin of the Glommen for the 185,335 dozens of wood delivered for floatage in 1880 a gross amount of 4,683,077 crowns, which sum, after deduction of 1,345,000 crowns for felling, sorting, carriage, and floatage, shows that the proprietors have been benefited to the extent of 3,337,077 crowns (£183,610.)

'The exports of wood from Sweden in 1879 were less satisfactory; and of about 890,000 dozens of old and new wood, 500,000 dozens only were floated and delivered.'

CHAPTER XX.

SHIPBUILDING AND SHIPPING.

IN Norway there is a great consumption of wood as fuel for domestic purposes and in manufactures; there is also a great consumption of it in buildings constructed entirely of timber, and in carpentry, in the manufacture of furniture, in the construction of railroads, and in the construction of carriages of various kinds. Surpassing all these in interest for foreigners, is the consumption of it in shipbuilding; but that more in reference to the shipping produced than the quantity of timber thus employed. From of old the Norsemen have been famous for their maritime enterprise. It may be that it was as sea-rovers that they found their way to Scandinavia. The course followed by the Lapps, and that subsequently followed by the Finns, can be traced from the east through Northern Russia; but it is not so with the Norsemen.

It is alleged that the Lofoden fishing boat of the present day is almost exactly of the same build as the war galleys of the ancient Vikings, in which they ravaged every shore of Europe from, the bleak and storm-beat coasts of Orkney and Shetland, to the sunny Isles of Greece. There is the same lofty prow, the same sheer, and the single lofty mast with its heavy square sail.

In a treatise on *Prehistoric Sweden*, by Oscar Montelius, it is stated that there was found in a peat-bog at Nydam, in Jutland, two large boats, accompanied by Roman coins of the second century, and numerous articles belonging to the first age of iron. They were clincher-built. The one of oak, the other of pine. They were not decked, and they terminated both before and aft in a point, were fitted only for oars, and showed no trace of a mast. The oaken boat was remarkable for the elegance and the flexibility of its outlines; it measured 24 metres, 80 feet, between the high points of the two stems; the greatest breadth was 3·50 metres, nearly twelve feet. It was propelled by means of 14 pairs of oars, exactly like those of our own day, the rudder was narrow, and had been attached to one of the sides of the boat.

A boat in most particulars resembling these, but which belonged to the age of iron, was found at Tune, near Frederikstad, in Norway. This had still preserved the remains of a mast.

A year or two ago an ancient Viking ship, which had been deposited in a grave mound at Sandeherred, in Norway, was discovered and disinterred for preservation in the Museum of Antiquities connected with the University of Christiania. Great importance was attached to it. While the operation of disinterment was being prosecuted, a correspondent of the *Hamburgh Correspondent*, writing from Bergen, said:—'A measurement made on the 14th May showed the length of the portion already excavated to be 71½ feet, and it is estimated that the total length will be about 75 feet. As regards the equipment of the ship, it is evident that, when deposited in the grave mound, it was as fully armed and equipped as when it lay ready to sail on a Viking expedition. So far as can be judged at the present stage of the process of excavation, all the appliances in use at the time for evolutions at sea are represented with quite remarkable completeness. The greater part of a mast in good condition remains; the entire length of the mast seems to have been about 22 feet. Remnants of sails and tackle are frequent, as also fragments or complete specimens of ships' utensils and divisions, the place and application of which it will be the difficult task of antiquarians to solve. Several wooden articles of a peculiar form have been found in excellent preservation, and executed with remarkable skill. That these articles, which are about one and a-half feet in length, were used as grooves through which the ropes passed is evident from the fact that their circular openings are completely worn by use—a proof, at the same time, that this ship must have frequently ploughed the sea before finding a last resting-place in the grave mound. How these pieces of wood, which are constructed of excellent oak, were fastened to the mast or the body of a ship, is a problem, the solution of which will be of great interest. If this point be successfully cleared up we will obtain a hitherto unexpected insight into the method in which a Norwegian Viking ship was handled during manœuvres in the eighth or ninth century. Among the other wooden implements are several spades almost entire. A fact of very great interest is the finding of many shields, or, speaking more correctly, remnants of shields; for of the wooden shield nothing has been preserved to us but the iron plate which strengthened its outer side. From the present position of these remnants it is evident that a great part of the inner side of the bulwarks formed hanging places for the shields of the crew. The distance between each shield is found to be almost exactly $1\frac{1}{2}$ feet, and it is probable that the number of the shields nearly corresponds with the number of fighting men on board. Of human remains only a few calcined bones have, as yet, been found, which seem to indicate that either the bodies of these latter, or of the persons buried on the spot had been burned. On the other hand, the skeletons of three horses have been discovered, two on the right and one on left side of the stern. It is possible that the opening of the grave itself, which is in the centre of the ship, may bring more human remains to light ; but the latter remains as yet untouched, the principal object having hitherto been to free the ship in its entire length and breadth from the surrounding masses of earth. After the excavation has been completed the ship will be drawn up to the surface.'

The roof of a cathedral church, built in what is popularly known as the Norman style of architecture, has been often likened to the over-arching boughs of a forest forming a long and lofty vista, with lesser, but like vistas, on either hand. In Norway I met with a different account of this: I found it likened to the upturned boat which was used in olden times to roof an earthen house, a sepulchre, or a temple—a representation in stone of an ancient boat so employed, commemorating with the luxury of advanced civilisation, wealth and power, memories of the childhood of the nation.

While an improved and increased shipping of late years witnessed in Norway is intimately connected with the later development of commerce in that country, as elsewhere, it is as manifestly an effect as it is a cause of that development; and in Norway, while it has facilitated the transport and exportation of timber, it has also increased the home consumption of this by the demand it has made for a supply of home-grown timber to be used in the construction of the vessels.

The Norwegians may be characterised as a maritime people; and in proportion to the population, the first place must be allotted to them in this respect.

At the end of 1875 the mercantile navy of Norway was composed of 7,814 vessels, of an aggregate tonnage of 1,419,300 tons English, with an aggregate crew of 60,281 men. This gives for every thousand of the population 781 tons, while at the same period there were for every thousand of the population of Britain 210 tons of British vessels. In the United States of America the proportion was 90, in Russia 10. Even in regard to tonnage alone Norway took the third place, being only surpassed by Great Britain and the United States of America.

By Dr Broch there are supplied tabulated statements of the number, tonnage, and crews of the Norwegian merchant navy in 1767, 1792, 1800, and every five years thereafter till 1850, and every year thereafter till 1876, giving the number of ships, the tonnage, the mean tonnage, the crew, the proportion of these per thousand tons, and the number of steamers, tonnage, crews, and horse-powers of these; and of the annual additions made and losses sustained by the mercantile navy in each year from 1861 till 1875, giving the number and tonnage of vessels built in Norway, of vessels purchased abroad, of vessels sold abroad, of vessels lost at sea, and vessels condemned; the means of each of these particulars in the successive semi-decades, and like particulars in both respects relative to steamers. He gives like tabulated statements of the number and tonnage of vessels entering Norwegian ports from abroad, loaded and in ballast, of vessels leaving Norwegian ports, loaded and in ballast, for foreign ports, and the number of each of these categories entering or leaving laden; the number of Norwegian vessels sailing between foreign ports, arriving and departing, and the number of these in cargo.

From the first of these tabulated statements it appears that the mercantile navy has doubled within the last preceding ten years, and quadrupled in the last twenty-three years. The increase dates from the opening of English ports in 1850, and from demands arising out of the Crimean war. But the second shows that while the increase in numbers was made both by construction and purchase, the great increase in proportionate tonnage was by purchase, the average tonnage of the vessels built being 228 tons, the average tonnage of those bought, 392 tons.

The average size of the vessels has been more than doubled within the last twenty years; in 1855 it was 83 tons, in 1875 180 tons.

The greater portion of the Norwegian merchant ships are sailing vessels, but the number of steamers is considerable.

The value of the mercantile navy of Norway amounted in

1850 to	. .	93 million francs.*
1868 to	. .	260 ,,
1874 to	. .	278 ,,
1875 to	. .	267 ,,

---

* In round numbers 25 francs may be reckoned equivalent to a guinea.

In general, the captain holds a share in the ship, and sometimes he is sole owner; and the greater part of the shipowners are old captains, well acquainted with what ships should be, and what ships should do.

Captains and mates of vessels sailing to foreign ports must undergo an examination in navigation, and must have served a certain time as seamen. There are 15 navigation schools maintained by the State, or by communes. From 1871 to 1876, upon an average 1432 candidates presented themselves for examination. The Norwegian seamen are generally skilful and well-behaved, and the country is proud of them. Many Norwegian seamen serve in foreign vessels, more especially those of Great Britain, and of the United States of America.

Dr Broch gives a tabulated statement of the annual consumption of victuals in the State hospital in Christiania, in the workmen's economic dining halls in Christiania, in the Royal navy, in the merchant navy, and in the army, in camp and in garrison. From this it appears that the diet in the mercantile navy is very satisfactory—the consumption of butter is astounding: it is given as 26 kogren per annum.

The census of 1865 gives the following as the numbers of persons mainly occupied with navigation:—

	Heads of Family.	Other Members of Family.	Domestics.
In Towns,	13,386	21,613	1,835
In Country Districts,	17,647	20,171	1,947
In whole Kingdom,	31,033	41,834	3,782

Or a total of 76,649 persons, equal to 4·5 per cent. of the entire population.

By Dr Broch there is given the number of strandings and of shipwrecks which have occurred on different parts of the coast; the total number of these; and the number

P

of men lost; with a classification of the vessels according to the national flag which they bore. From these it appears that the mean annual loss from 1871 to 1875 was 2·37 per cent. of the vessels, and 3·05 per cent. of the tonnage. On an average 176 men were lost annually by shipwreck; in 1873, 1874, 1875, about 300 men, it was calculated, died of disease.

There are numerous maritime insurance companies in Norway. Formerly, say from 1814 to 1837, assurance was generally effected in Hamburg. In 1875 there were in Norway 13 assurance companies, insuring that year against loss 282 million francs, on premiums amounting to 2,780,000 francs.

The coast is well lighted, according to most approved methods, throughout its entire extent. In the end of 1876 there were 120 lighthouses, and one light-ship, many of them of a high-class, and constructed according to the system of Freesnal. Of these 108 were maintained by the State, built at an expense of 5,700,000 francs. The others are port lights, maintained by communes. The expense of lighting the coast and of signals was in 1875 808,600 francs.

Pilotage was first established by law in 1720. Sundry changes have from time to time been made in the laws relative to it. The fundamental regulations now in force are embodied in the law of 17th June 1869. These prescribe the organisation and the tariff. Pilotage is compulsory only on vessels coming from or going to foreign ports, and vessels of above 104 tons going to or returning from the fishing. The prohibition of employing others than members of the staff, which was previously in force, was then abolished; and the obligation to employ these in their turn in going out, and to employ the first who boards on entering, was modified so far that such must be *paid*, but they need not be employed.

A fund in aid of pilots was organised in 1805, lending money free of interest for equipping pilot boats, and supporting the aged and infirm, and widows and children.

The Norwegian pilots are famous for their skill, and energy, and daring; but every year some are lost. In the fifteen years 1861-1875, 126 pilots were drowned, of which 98 perished in the service, and 26 at sea, but not at the time so engaged.

At several localities there are stationed life-boats and safety rockets, the latter being in some places necessary, the stony coast being such that no life-boat could reach the shore in a storm. The annual budget for expense of maintaining these stations amounts to 1700 francs. In the twenty-two years subsequent to 1855, when the stations were organised, safety rockets have been employed in 11 cases of shipwreck, and they have saved 69 men.

For all of this information I am indebted to the report by Dr Broch.

CHAPTER XXI.

FOREST DEVASTATION.

NOTWITHSTANDING the great recuperative power manifested by Norwegian forests, of which mention has been made, they are, under the excessive drain made upon them in districts favourably situated for the prosecution of trade in timber, being greatly impoverished. Of this there are indications in more than one of the official reports mentioned in connection with the information given in a previous chapter relative to the geographical distribution of different kinds of trees in Norway. Amongst these is a report on the condition of the forests in Romsdal county, which lies between the Dovrefjeld and the sea, along the coast to the south of Drontheim fiord, where the shore trends to the south-west, passing the town giving its name to the county. The report is by Johannes Schioetz, LL.B., Forest-Assistant, and was issued by the Forest Section of the Department of the interior in 1871.

In this it is reported in regard to the amount of forest in the County of Romsdal:—

'Large forests cannot be expected in this county owing to the natural formation, which is mountainous, and greatly intersected by fiords; moreover, the narrow valleys are cultivated, and the mountain sides often nearly bare rock. The inhabitants are often in want of fuel: turf also being rare. Even in the lower districts the mountains rise above the forest line, and on the coast the inhabitants and sea-breezes have almost completely destroyed all the forests which existed there in earlier times. Only on the islands in Nordmore are there some important exceptions.

'The forest line does not lie high, and depends not so much on temperature as on sheltered position. Among

FOREST DEVASTATION. 213

the fiords firs grow commonly at 1000 feet above the sea level, and in very sheltered spots, even so high as 2,500 or 2,800 feet, but all these are small and stunted. Trees grow best at about 300 feet above the sea level, where the soil seems to be best. All over the district it is very sandy.

'Coniferous forests are exclusively formed of fir, and are very unequally divided among the three bailiwicks, being in inverse proportion to the mountainous formation. The higher the mountains the smaller the woods, until they vanish altogether.

'While in Nordmore there is a good deal of wood, sufficient for home wants, with a small surplus for exportation, in Romsdal there is much less, and in Soudmore the fir is seldom seen, and, with the exception of some good forests of foliage trees, there is a great scarcity of timber, which must either be brought from a distance or bought at a ruinous price on the spot. Turf is known and used, but not so extensively as might be expected, the people preferring to destroy the last remnants of the forests rather than adopt anything new. In all the three bailiwicks the forests have deteriorated for the last ten or twenty years on account of the increasing population and the high price given for timber.

'The fir woods are badly managed, being thinned in the usual way, and consequently there are many trees stunted for want of light and room at an early period of their growth. The fellings are made without judgment, and too frequently. It is only lately that small timber was needed for various purposes, and now the trees have no chance of growing to any size, and are already vanishing in a very sad manner, fortunately heath burning, so destructive in other places, is here unknown.

'There seem to be no communal forests in this county, although in the public archives I found notices of some in Surendalen and Surrdalen, but I had not time to make inquiries. In the first district I rather think they have been allotted to private individuals, although there may still exist one or two of small extent.

'Of different trees, Pinus sylvestris, Abies excelsa, and Sarix Europea are found in pleasure grounds; also Abies pectinata, first planted by a Scotchman. Betula alba is the most common tree next to the fir. We have Alnus incana, Alnus glutinosa. Corylus avelana is much used for making barrel hoops. Populus tremula is found everywhere, but not in great numbers. Ash grows wild, but is always sadly misused for the sake of the foliage for the cattle. Mountain ash is common, especially on the islands, where it is mixed with birch. Oak is found wild in one or two places, but there only like a shrub. Besides these, there are the following, of no economic value:—Salix caprea, bird-cherry, many willows, Sorbus area, holly, white thorn, and crab apple; lime trees are also found in one or two places.

'*Nordmore Bailiwick.*—The greatest extent of wood is found here; formerly it must have been much greater, which is seen in the large size of the beams in old houses, many of which are from 8 to 12 ells long, and five quarters in breadth. Long ago this was a secluded district, and prices were not so high as to tempt the people to cut down the woods. Most of the inhabitants were engaged in seafaring pursuits. Even yet the sawing up is done by the hand, and of course not on a large scale. It also seems as if the fiord and coast population were economical in their consumption of the timber, especially with regard to young trees Had they been as extravagant as their neighbours the woods would have altogether disappeared, the climate and soil being both bad. As it is, there is a great change for the worse within the last ten years. The population is rapidly increasing, and the sale of Government lands is hastening the work of destruction. Many of the farmers (*bonderne*), are awake to the evil, and wish there were some regular system of forest management. Nowhere have I seen the pernicious practice of leaf stripping so common, even in the path of the avalanches, which, as the forests disappear, are becoming more frequent and formidable; one killed thirty men in one night.

FOREST DEVASTATION. 215

This also causes a great want of fuel, particularly in the drier districts, where turf is unknown.

'The climate also suffers in this extremely dry rocky district. In winter there is no shelter from the icy winds, and everything is scorched in summer, unless the weather be very wet.' There follow local statistics which I do not think any of my readers would find interesting.

'*Romsdal Bailiwick.*—The amount of wood is almost the same as in Nordmore, with only this difference—that Romsdal has always possessed fewer fir woods, and even these are rapidly disappearing. In the seventeenth and eighteenth centuries Britain imported largely from this district; and its near neighbourhood to scantly wooded Soudmore, has occasioned a further drain on its resousces. The inhabitants say that the day of the fir is past, that it will not now thrive. This is a mistake. Good management is all that is needed, both climate and shelter being pretty good, and many trees rare in Nordmore are here quite common. The people are economical in their own use of the timber, though using unhewn timber for their houses, or at most, with only two planks cut off. Here also the climate has changed for the worse, wherever the woods have been cut down, and the people are themselves aware of this.' There follow statistics of the several parishes.

'*Soudmore Bailiwick.*—The whole of this district is very mountainous, some of the peaks being 4000 feet high. There is only a small area fit for the growth of timber, and even this is much exposed to both sea-storms and avalanches. Sheep and cattle go at large, and greatly add to the destruction of the young trees. All these circumstances, combined with communal privileges, caused the fir woods to vanish long ago. In fact, the scarcity of trees is by the country people said to have been the work of enemies in the olden time; but I do not believe this. The fragments of charcoal still to be found scattered on the fields may have either been purposely manufactured, or the trees may have been accidentally burned.'

'The foregoing plainly shows the scarcity of wood in Soudmore, the people being obliged to import largely. Matters have come to such a pass that Government ought to interfere to prevent further mischief. To further this end I offer the following advice:—

'1st.—That all woods attached to clerical or other offices should be taken in charge by the Government. Their extent is small; but doing this would spread information, and awaken an interest in forest science, especially in Nordmore, where the people are intelligent enough to see how profitable the forests might be made. I found it very encouraging that the farmers voluntarily accompanied me to the woods for the sake of instruction, although it was in the midst of harvest. They expressed sorrow and wonder that the Government had not interfered earlier.

'2nd.—The purchase of forests on a great scale by Government would also be a profitable investment, but in Soudermore there are some obstacles to this. The land is much subdivided and expensive; and the rights of pasturage and communism would be an almost insuperable obstacle to the purchase of great continuous stretches. It is much to be regretted that the Crown lands were sold in this district.

'3rd.—All Government employés should be bound to assist and instruct the inhabitants in the best modes of forest culture.

'4th.—Great stretches should be planted. The climate and soil are pretty good; shelter at first is all that is required; but here again we are met with the obstacles of rights of pasturage and communal privileges.

'5th.—Scientific planting and development should be steadily carried on. Something has already been done in this direction by Storthingsmand Aarflot, and Judge Thamb, by a distribution of seed; but with small result, except in attracting attention to the subject, which it has now done in some measure.'

As is the case with this so is it with others of the reports from forest officials, they all ring the changes on the same topic, the rapid disappearance of the forests, especially those of fir. They report that Government must do the work itself; nothing or little can be expected from private proprietors: only one of the latter is spoken of approvingly as making a movement in the right direction.

The rights of pasturage are reported to be a great obstacle, as the young trees must be fenced in order to protect them from sheep, which in Norway seem to be very agile in that particular part of the west coast, needing a higher fence than usual. The expense of fencing falls of course on the improving proprietor. One writer says Government should interfere on this point, for nothing will be done so long as sheep can roam at large. Another brings the same complaint against cattle.

It is recommended that the heath under the trees should be cut down when there is a good seed year, as it chokes the young plants. Another suggests that all woods attached to clerical or other offices should be resumed by Government; their extent is small, but the operations would give instructions to proprietors in the neighbourhood. In some places land is cheap, and could be bought to advantage, in others it is very much subdivided, and difficult to obtain.

All write gloomily, and the only hope is in the Government buying and replanting large stretches, and limiting the rights of pasturage.

It is recommended further, that Government officials should be required to assist all who wish their help in making attempts at forest culture. In general the people only care for money, and are careless of the future, with, of course, cheering exceptions. One English company, it is reported, has been hewing down at a great rate.

Something similar may be said of the reports on the economical condition of the kingdom, made by the prefects of nineteen prefectures, and a report of the same to the king by the Department of the Interior, with copies of administrative instructions, and forms of tabulated returns required, and copies of tabulated returns received in regard to agriculture and cattle, including meteorological observations, in regard to fishing and the chase, in regard to industrial occupations and products, in regard to commerce, navigation, and roads, and in regard to finance and financial conditions for the years 1861-1865 ; and there are given official reports on the economic condition of the several prefectures for the years 1866-1870.

In the report on the forest condition of Lister and Mandal—a report made to the Department of the Interior by Forest-Assistant Aars, published in successive numbers of the *Christiansand Stiftsavis* in the latter months of 1870, the whole series is charged with most interesting minute local descriptions; but the burden of the whole is complaint of the disappearance of the forests. The first district mentioned, Vauso and Herred, is described as subject to inundations of drift sand ; and the planting of the ground with trees, and the covering of the sand with sea-weed are suggested as remedial measures, the adoption of which is urgently called for. Of the *Praestergjeld*, or parish of Fjotland, the clergyman writes that 'things are bad and every day becoming worse; as time goes on every stick growing will be converted into money, and then —— ?' From another district the clergyman writes that there is no hope unless the Government come forward and purchase the remaining forests. From a third district the clergyman, after pouring out a long Jeremiad of lamentations, winds up with a like suggestion as the only measure likely to prevent utter devastation ; and from the *Praestergjeld*, or parish of Siredalens, the report is that the destruction of forests has been carried so far that even the interposition of the Government could effect nothing in preventing desolation, as some places must ere long be almost uninhabitable.

The report was furnished to me by the Government; it is in exact accordance with what I have seen of the results of reckless felling elsewhere.

CHAPTER XXII.

REMEDIAL MEASURES.

In Norway it was reported in 1882 that there were 147,000 farms, 131,000 of which were farmed by their owners, and to the recklessness with which these treat their forests was attributed much of the impoverishment manifested by these.

In view of the waste which has been going on, and of the extent to which lands comparatively unproducted might be utilised profitably by planting them with trees, it was deemed expedient that more attention should be given to this in the education, instruction, and training of students at the National School of Agriculture and Rural Economy at Aas, than had been, and was being, done. And advantage was taken of a re-organisation of this school in 1871 to secure the accomplishment of this. And something is being done to secure at the same time the extension, as well as the conservation and improved exploitation of the forest.

The following is a translation of a report on forest cultivation at Aas, by Mr H. Fougner, which has been sent to me:—' Forest cultivation was commenced at Aas in the year 1868, when, as a first attempt, trees were planted on two small pieces of woodland close to the forest on the Söraas Hill. On one of these about $\frac{1}{2}$ maal (1 maal = 02,363 acres) were planted different kinds of foreign fir trees—as Pinus austriaca, Pinus strobus, Pinus cembra, Pinus montana, Abies alba, Abies pectinata, and Larix Europæa; on the other piece about $\frac{3}{4}$ maal were planted mostly deciduous trees, more especially elm, maple, and ash, mixed with small groups of Pinus austriaca, and Pinus strobus. Both these plantations, after having been a little filled up in the following year on account of damage caused partly by mice and partly by frost, have since then succeeded well, and they show at present on the whole a good, and in some parts even a very good, growth. In the spring of 1870 some maple and a little ash were planted in groups on more exposed places within the same woodland (about $\frac{1}{2}$ maal). A part of this plantation has, however, been damaged by frost and by mice, which have gnawed the bark off the trees; and it does not look so well as the plantation made in 1868; within the last two years, however, it has shown a better growth, and it seems to be thriving. In 1871 attempts were made to rear forests by sowing. The upper or easter forest at Aas, on the hill between Frydenhaug and the workmen's lodging house, a piece of about $3\frac{1}{2}$ maal, was sowed with fir seed mixed with larch seed. This succeeded well on the whole; and with the exception of a few spots, where the thick grass had prevented the growth of the small plants, this sowing has at present a good appearance. In the lower or western forest, between the high road and the neighbouring farm, Kvestad, about 1 maal was also sown with fir seed mixed with larch seed; this, however, did not succeed so well, on account of there being here another kind of soil. The sowing of larch seed was a failure, and therefore we find here more bare spots, which will have to be afterwards filled up. Next to this sowing, about 2 maal were experimentally sown in the same spring with oak and ash, the experiment was, however, a failure, the plantation, situated on a low level, close to the high road, having since suffered so much from frost that it has had for the greatest part to be cleared and replanted with pine and some Betula oderata; the replanting, which was commenced in 1873, and continued in 1875, looks well. All of these experimental plantations in 1868-1871 (all together about $8\frac{1}{2}$ maal) were executed exclusively by the pupils of the school, although forest cultivation at that time was not not among the ordinary subjects of teaching which were prescribed.

'When the Agricultural School at Aas was re-organised in 1871, forest cultivation, sowing as well as planting of woods became, in accordance with the rules of the school, an obligatory work, and therefore cultivations of wood entered a new stage. Up to this time the work done had been considered merely as experiments, and did not belong to the ordinary instruction; but in the sequel special importance had to be attached thereto, and the pupils had to be instructed as well in sowing as in planting. Thus the forest cultivation came into more intimate connection with the forest belonging to the school, forming a regular and constant contribution to the renewal of woods; the works will also serve as experiments, from which may be drawn knowledge relative to artificial cultivation of woods under ordinary favourable circumstances in the easterly low districts, and such knowledge may gradually become valued and utilised by others; finally, if the cultivations should succeed, the example would give an impulse, better care being taken of the woods after growth in general, and to the raising of thicker wood, when the soil would be turned to more use than has been the case up to this time in the neighbouring districts. Pursuant to these reflections it may be said that the present forest cultivation has to serve three different ends—in the first place as experiences for the pupils of the school; in the second place as forming a continuous contribution to the cultivation of the forest belonging to the school farm; and, finally, as an example relative to the raising of woods of normal denseness. These three ends have always been kept in view, when the forest cultivations were commenced and executed. Therefore as a rule the cheapest system of cultivation has been employed, and only, where it was absolutely necessary has a more expensive method been followed. Regarding the qualities of trees used special importance has been attached to the raising of as many different kinds of trees as possible, and to forming as heterogeneous—partly pure, partly mixed—stocks as practicable, in order to procure the greatest possible material from which to judge for the future what would be most suitable and best to choose for the purpose of forest cultivation under circumstances similar to those at Aas. This, however, entails, what it is difficult to avoid, that some pieces of the cultivated woodland do not succeed so well as others, and that some small bare spots have to be replanted; this inconvenience will, however, be removed by more experience.

'Further, it must be noticed that the present forest cultivations, on account of their serving in the first place as experiments for the pupils of the school, can consequently never be expected to be executed with the same exactness and prospect of a successful result as if workmen of experience could be employed; the pupils having only when they have finished their instructions in forest cultivation in each course, acquired the experience necessary, and such knowledge of matters as that would be required from an ordinary trained planter. The practising foresters in Germany are wont, in a well-known phrase, to talk with a certain disrespect of "academical plantings": plantings executed by the students at the agricultural academies; and the same thing might consequently with some right be used about the plantings executed by the pupils here. This point should therefore be well considered when criticising the cultivations; and it must not be forgotten that these cultivations are, and must be considered as, works of inexperienced pupils; on which account the ordinary plantings here will also always have a comparatively greater extent than would be the case under other circumstances.'

After these remarks, which were considered necessary to draw attention to the forest cultivations executed subsequent to the same becoming an obligatory subject of instruction, the report proceeds:—

'In the year 1872 about 3 maal on the hill between Frydenhaug and the lodging-house for the workmen, near to the sowing of 1871, were sowed with fir seed, mixed with a little larch seed. The seed came up all right, and it looks well. On the space between the high road and Kvestad, about 12 maal, were sown with fir seed, mostly fir and pine, and some larch, mixed together in different proportions. The seed sprouted well, but the small plants suffered the first year partly from wetness, and the following winter partly from frost, later also a little from the strong growth of grass. In the spring of 1873 a fire broke out in the sowing, through which there is a footpath. The fire originated probably from carelessness with matches of passers by, and about ½ maal of the sowing was destroyed, but it was resown the same year. Later it suffered also a little from unlawful pasturing, and last year from the military manœuvres when it served as battlefield for a detachment of the infantry attacking the position at Aas. This sowing shows, therefore, at present a rough and less favourable appearance, even less favourable than it deserves; and there are some bare spots which will have to be filled when the ordinary replantings commence. It is worth while to notice that there is a distinct difference relative to the qualities of trees, the fir having succeeded best, next the pine, the larch appearing poorer. The soil may be characterised as fir wood land with some swampy spots, and now and then a little harder clay underground where planting would have been more effective, if there had been a supply of suitable plants, which was not the case at that time. This year a planting of about 2 maal of foliaceous trees was executed on the space close to the high road, and nearer to Brisnenid, consisting of oak, ash and birch, in groups, mixed a little with elm and maple. The planting has up to this time succeeded tolerably well, although some of the ash and elm trees have been damaged by frost, especially in the strong winter 1874-75, whilst on the other side the oak plants have come through this inconvenience much better. Besides, some foreign firs, oak, and larch were planted in the same year's planting on some small spots a little further from the high road. The forest cultivation in 1872 thus embraces together a little more than 17 maal.

'In the year 1873, on the hill between Frydenhaug and

the lodging-house for the workmen, close to the last year's sowing, about 1 maal was sowed fir seed, mixed with larch seed. The sowing has a good appearance, but the plants are yet too small to warrant any decisive opinion as to this. On the hilly ground at Aake about 12 maal were sown with mixed fir seed, consisting mostly of fir, a little less pine, and a little larch. The seed came out well, but suffered the following year from the sharp and dry winds on the more exposed places, and later on from the thick grass on other spots; yet the sowing on the whole appears well, although it is yet too early to form any decided opinion.

'In the year 1874, on the hilly ground between Frydenhaug and the workmen's house, a piece about 8 maal were sowed with mixed fir seed, in the same proportion as regards the mixing as on the Aake hill. The seed did not come up as evenly as was desirable, because of the soil in this place being very much exposed through sloping down southward, and its suffering very sorely from drought; it is yet too soon, however, to give any reliable opinion. At the foot of the same hilly ground, below the sowings, about $7\frac{1}{2}$ maal were planted with pine. This piece consists for the most part of an abandoned gravel pit, with very uneven surface, partly covered with stone, partly with pools, which made the planting very difficult. The plants however appear to thrive tolerably well. The whole area planted in 1874 amounts thus to $12\frac{1}{2}$ maal.

'In the year 1875 the rest of the hilly ground between Frydenhaugh and the workmen's lodge, about 9 maal, situated nearest to the lodging-house was this year sowed with a similar mixing of fir seed, which succeeded very well. The planting of pine at the foot of the hill was continued for about $1\frac{1}{2}$ maal towards the lodging-house. In this plantation were mixed some foreign fir trees. At present there is every appearance of a good result. The forest cultivation in 1875 comprised altogether $10\frac{3}{4}$ maal.

'The forest cultivation here from 1868-75 extended to about 64 maal in all.

REMEDIAL MEASURES. 225

'As a general remark it may be said that the forest cultivation here,' as is so often the case with plantings of trees in this country, 'has also suffered and been damaged by mischievous and evil-minded persons. Repeatedly trees have been broken and the bark fleeced off; sometimes even the trees have been torn up by the roots and thrown on the field, consequently it was necessary to put up warnings against violence and damaging the forest plantings, and this has led to some improvement in this respect.

'In conclusion, I desire to state some results gained by experience in the eight years' forest cultivation which has been carried on at this place. The fir seed sowed has given better results than pine seed, and this again better than larch seed.

'In sowing the different kinds of seed they must be mixed with different kinds of common "rudesod;" and in the mixing fir must be the predominating sort, as a means of avoiding bare spots and obtaining more even growth.

'In plantings of pine, the plants, four years old, which have been transplanted, give a better result than younger ones, in ordinary hole plantings.

'Larch seed ought to be sown as early in the spring as possible on account of its early budding.

'In planting the better sorts of foliaceous trees, such as oak, elm, maple, beach, &c., it is under similar favourable circumstances safer to choose hilly than low land, thus avoiding damage by frost. *Dunbeach* (Betula oderata) can, however, without any risk, be planted in low-lying places and in damp moist land.'

With this report there was supplied to me a document issued by the Budget Committee in 1872 entitulated *Ved Kommende Skovsagen*, in one of which are described a number of estates on forest lands about to be purchased on behalf of Government; and in another is given the staff of forest officers, and the locations of these in different districts. This was as follows:—

1. Prefectures of Akershus and Hedenmark—
    1 Forst-Meister stationed in Christiania, and one assistant at Hamar.

2. Prefecture of Christiania—
    1 Forst-Meister at Lillehammer.
    1 Assistant in Valders and Thoten.
    1 Assistant for Almindingerne in Vaage and Lorn.
    1 Assistant for Lesje, Dovre, Foldalen, and Udmaalingsskovne, at Fjeldstuerne, stationed at Dovre.

3. Drontheim Forest District.
    1 Forst-Meister at Levanger.
    1 Assistant in the Prefecture of Soudre Drontheim.
    1 Assistant in Stor and Vœrdals, Fogderi.
    1 Assistant in Resten of Inderœn.
    1 Assistant in Namdalen.

4. The Prefecture of Nordland—
    1 Forst-Meister in the Paræstergjeld or Parish of Skerstad.
    1 Assistant in Beieren.
    1 Assistant in Saltdalen.
    1 Assistant in Ofoten and Hommerœ.

5. Prefecture of Tromsœ and Finmark—
    1 Forst-Meister at Tromsœ.
    1 Assistant in Malangens and the Maalselv District.
    1 Assistant at Alten.
    1 Assistant at Tauen.

6. Vestland District—namely, Prefectures of Lister and Mandal, Stavanger, and Southern and Northern Berganhus—
    1 Forst-Meister in Sognedal.
    1 Assistant in Bergen.

Together with these are the necessary functionaries and Forest-Warders and labourers. In Varanger a *Reiverfervalton*, and elsewhere other officials. There is also an estimate of the revenue and the outlay.

THE END.

www.ingramcontent.com/pod-product-compliance
Lightning Source LLC
Chambersburg PA
CBHW021817230426
43669CB00008B/776